TAPPING THE ZERO-POINT ENERGY

Moray B. King

Author of *Quest for Zero-Point Energy*

How "free energy" and "anti-gravity" might be possible with today's physics

The **New Science Series**:
- •THE TIME TRAVEL HANDBOOK
- •THE FREE ENERGY DEVICE HANDBOOK
- •THE FANTASTIC INVENTIONS OF NIKOLA TESLA
- •THE ANTI-GRAVITY HANDBOOK
- •ANTI-GRAVITY & THE WORLD GRID
- •ANTI-GRAVITY & THE UNIFIED FIELD
- •ETHER TECHNOLOGY
- •THE ENERGY GRID
- •THE BRIDGE TO INFINITY
- •THE HARMONIC CONQUEST OF SPACE
- •VIMANA AIRCRAFT OF ANCIENT INDIA & ATLANTIS
- •UFOS & ANTI-GRAVITY: Piece For a Jig-Saw
- •THE COSMIC MATRIX: Piece For a Jig-Saw, Part II
- •TAPPING THE ZERO-POINT ENERGY
- •QUEST FOR ZERO-POINT ENERGY

The **Mystic Traveller Series:**
- •IN SECRET TIBET by Theodore Illion (1937)
- •DARKNESS OVER TIBET by Theodore Illion (1938)
- •IN SECRET MONGOLIA by Henning Haslund (1934)
- •MEN AND GODS IN MONGOLIA by Henning Haslund (1935)
- •MYSTERY CITIES OF THE MAYA by Thomas Gann (1925)
- •THE MYSTERY OF EASTER ISLAND by Katherine Routledge (1919)
- •SECRET CITIES OF OLD SOUTH AMERICA by Harold Wilkins (1952)

The **Lost Cities Series**:
- •LOST CITIES OF ATLANTIS, ANCIENT EUROPE & THE MEDITERRANEAN
- •LOST CITIES OF NORTH & CENTRAL AMERICA
- •LOST CITIES & ANCIENT MYSTERIES OF SOUTH AMERICA
- •LOST CITIES OF ANCIENT LEMURIA & THE PACIFIC
- •LOST CITIES & ANCIENT MYSTERIES OF AFRICA & ARABIA
- •LOST CITIES OF CHINA, CENTRAL ASIA & INDIA

The **Atlantis Reprint Series:**
- •THE HISTORY OF ATLANTIS by Lewis Spence (1926)
- •ATLANTIS IN SPAIN by Elena Whishaw (1929)
- •RIDDLE OF THE PACIFIC by John MacMillan Brown (1924)
- •THE SHADOW OF ATLANTIS by Col. A. Braghine (1940)
- •ATLANTIS MOTHER OF EMPIRES by R. Stacy-Judd (1939)

TAPPING THE ZERO-POINT ENERGY

Adventures Unlimited Press

Tapping the Zero-Point Energy
by Moray B. King

ISBN: 0-931882-00-2

Published by

Adventures Unlimited Press
One Adventure Place
Kempton, Illinois 60946

Printed in the United States of America

Published in association with
Paraclete Publishing
P.O. Box 859
Provo, UT 84603

www.adventuresunlimitedpress.com
www.adventuresunlimited.nl

Dedicated to all inventors who freely give of themselves so that our planetary Being may be uplifted.

TABLE OF CONTENTS

FOREWARD

I have not always believed it was possible to tap energy from the fabric of space. As a Ph.D. graduate student in systems engineering, I held the standard view shared by most scientists and engineers. I knew the vacuum was empty because Einstein's theory of relativity did not need an ether. It was in the summer of 1974 that I had the misfortune of reading *Beyond Earth,* a book about UFO's. I picked it up just for fun, to read like science fiction. But what impressed me were the witnesses. Many were credible, such as airline pilots and police, who had everything to lose by reporting what they saw. The observed flying craft could undergo incredible acceleration and hairpin turns. They definitely appeared to exhibit an antigravity, or perhaps more accurately, an artificial gravity propulsion. This prompted me to ask the question: Is antigravity possible, or is artificial gravity possible? Since I wished to use this investigation as a thesis topic, I imposed the following constraint: I could only use sources in the standard physics literature and journals. In other words, did our physics today contain the principles that allow artificial gravity?

At this point I studied the "bottom line" theory of gravity, Einstein's theory of general relativity. From it I learned that gravity was a curvature in the space-time metric induced by the stress-energy tensor. This tensor could be comprised of mass or energy for they are equivalenced by $E=mc^2$. To achieve levitation on the surface of the earth required an enormous energy whose mass equivalent was 10^{12} grams. Things were not looking hopeful if we had to provide all this energy. Then, however, I discovered in the last two chapters of Misner, Thorne and Wheeler's *Gravitation* that in quantum mechanics there existed an all-pervading energy imbedded in the fabric of space consisting of fluctuations of electricity. It was called the zero-point energy. Zero-point refers to absolute zero degrees Kelvin. Wheeler's *Geometrodynamics* showed that the energy density was enormous: 10^{93}grams/cm^3. Quantum mechanics showed that this energy was constantly interacting with matter and the elementary particles in what is called vacuum polarization. If only a small amount of this energy could be made coherent in a statistical sense, then not only could artificial gravity be induced, but the energy could be tapped as a source as well.

At this point I asked my professors if it was possible to tap the zero-point energy. I was surprised to find most did not know this energy existed. Those who did, replied it could not be tapped because the action of this energy is random, and random things must forever remain random. This is the law of entropy, the second law of thermodynamics. Things were looking bleak until I discovered the work of Ilya Prigogine, who won the 1977 Nobel Prize in chemistry for identifying under what conditions a turbulent system may evolve from chaos to self-organization. These conditions were stated in general system terms, and the published theories of the zero-point energy could fulfill these conditions!

To build a theoretical case for tapping the zero-point energy requires merging two areas of physics: Theories of system self-organization and theories of the zero-point energy. I have found that most scientists are specialists and are generally unfamiliar with both theoretical areas. Those that are agree that a speculative case can be built, but it requires an experiment to prove it. To this I whole-heartedly agree.

To encourage experimental research, I have given a series of talks and papers over the last fourteen years for engineers and inventors. As an engineer, I was amazed at the marvelous possibilities arising from the theoretical constructs of modern physics. Each presentation was intended to stand alone and introduce the concepts in physics that allowed the possibility for a new technology. Thus as a collection, the chapters will contain a certain amount of redundancy. On the other hand this allows reading the book in any order. Let your intuition guide you in this.

I would like to express my thanks and deep appreciation to the following individuals who donated their time and talents in helping to produce this book: David Faust, Carl Rhoades, Andrea Powell, Dan Olson, and Rita Fryer. I would also like to thank the International Tesla Society and the United States Psychotronics Association for the opportunities to present this work.

TAPPING THE ZERO-POINT ENERGY

May 1978

ABSTRACT

Quantum mechanics claims the vacuum consists of fluctuating energy. Recent advances in theories of the zero-point energy and nonlinear thermodynamics open the possibility of cohering this energy. This could be verified by repeatably producing ball lightning in the laboratory.

INTRODUCTION

Modern physics may allow the possibility of tapping energy directly out of the fabric of space. While studying physics as a graduate student, I ran into a most interesting set of papers.[1-8] They stated that totally empty space was filled with fluctuating energy. As an engineer caught in an energy crisis, two questions arose. Was the energy *really* there and, if so, could it be tapped as a source? I talked with many scientists on this matter and discovered a remarkable thing: Most did not believe this energy existed.

However, I did run into some physicists who were already familiar with the concept. When I asked them, "Why can't this energy be tapped?" the reply was, "It would violate the second law of thermodynamics, the law of entropy. Random fluctuations must forever remain random." To them, there was no way to influence this energy.

Then I discovered the work of Dr. Timothy Boyer[5] who showed that matter influenced this fluctuating energy. And recently, I discovered the work of Dr. Ilya Prigogine,[9,10] the 1977 Nobel Prize winner in chemistry, who expanded the second law of thermodynamics to show how certain systems may evolve from randomness toward order. Combining their work opens up the possibility, in principle, that the fluctuating energy of space may be cohered into a source. It requires new physics, physics of the 1970's, to open the theoretical door. I predict an experiment will come through that door. The repeatable production of ball lightning in the laboratory may prove it is possible to tap the zero-point energy.

THE ETHER

The notion of a plenum embedded in the fabric of empty space is not new to science. During the 18th and 19th centuries, the ether was considered the all-pervading medium which would sustain light waves. At the turn of the century, Michelson and Morley attempted to detect the ether wind. Such a wind would be present if the earth were moving relative to a static, material ether. When Michelson and Morley failed to detect such a wind, Einstein used this result to verify his first postulate of relativity known as Lorentz invariance. This states that all observers moving at a constant velocity will experience the same laws of physics. Failure to detect the ether wind resulted in the general belief that no ether exists. Note the Michelson and Morley experiment only rules out a static ether; it is perfectly viable to have a Lorentz invariant ether.[1,2,46] Nikola Tesla[32] inventor of the alternating current generator, designed his devices based on a belief in an ether, and he argued quite vocally with the scientific community on this matter. When relativity theory became popular, Tesla's later designs were discredited. The scientific community and Tesla could have resolved their differences simply by considering a Lorentz invariant ether model. Then, they both would have been right.

A special class of ether theories describe space as a sea of fluctuating energy. These theories are significant because quantum physics predicts that vacuum fluctuations exist and gives them the name *zero-point energy*. The words "zero-point" refer to the fact that these fluctuations persist even at zero degrees Kelvin. There are many descriptions of the vacuum energy in the physics literature. In the 1930's Dirac[3] derived the idea that the vacuum consisted of a virtual sea of fluctuating electron-positron pairs. The discovery of the positron a few years later popularized Dirac's theory and the concept of vacuum polarization entered physics: Electric fields can affect the sea of fluctuating virtual charges.

By applying the theory of general relativity to the zero-point energy, Wheeler[4] derives a bizarre view of the fabric of space. The large energy densities of the zero-point fluctuations cause space to pinch in a manner similar to the formation of black holes. Wheeler views the vacuum as a fluctuating sea of mini-black holes and mini-white holes that pass electric flux through hyperspace channels he calls *wormholes*. This fluctuating sea, called the *quantum foam,* allows multiple connectivity: Distant objects in space can be instantaneously connected. Since the connections are random and constantly fluctuating, the theory maintains macroscopic causality. But if these connections could be technologically controlled, the possibility for teleportation would arise. Then the only way to maintain causality would be to accept Everett's *Many Worlds Interpretation of Quantum Mechanics*[26] where an infinite number of universes exist parallel to our own!

Another theory of the zero-point energy that has achieved successful quantitative results, is the random elec-

Table 1: A New Viewpoint

Quantum Effects Arise from a Matter, Zero-Point Energy Interaction

Quantum Event	Qualitative Explanation
Photon	Resonant absorption. Wave chopping occurs at detector
Quantum Eigenstates	Jump resonances of a nonlinear system
Ground State Stability	Zero-point radiation pressure balances Coulomb attraction.
Photoelectric Effect, Comptom Effect	See Scully and Sargent[6]
Blackbody Radiation	See Boyer[5]
Uncertainty Principle	Zero-point energy produces Brownian motion
Spontaneous Emission	Zero-point energy absorption
Pair Production	Soliton formation
Tunneling, EPR Paradox, Bell's Theorem, Nonlocal Connections[24,25]	Wheeler's "wormholes"[4] Hyperspace connections[27]
Infinite Self-Energies	Infinite zero-point energy flux implies higher dimensions of space
Renormalization	(Net energy of coherence) = (Infinite self-energy) - (Infinite incoherent zero-point energy)
Wave-Particle Duality	Waves are cohered zero-point energy; particles are solitons

trodynamics of Boyer.[5] He derived the spectral characteristics of the zero-point energy by assuming a Lorentz invariant spectrum. He thus mathematically described how the zero-point energy oscillates in its interaction with matter. Recently Boyer has introduced a new viewpoint to physics: Quantum effects arise because matter interacts and influences the fluctuating zero-point energy (Table 1). Note here the photon need not exist as a particle.[6] It only appears as such due to the wave chopping nature of the detection or absorption process. The mathematics needed to quantitatively support this viewpoint is overwhelmingly difficult. Some success has been achieved, but it will require new techniques in nonlinear analysis to complete. The appeal of this viewpoint is that it does not require special quantum postulates, thus it gives a more unifying view of the universe.

Despite the difficulty (Table 2), experiments have detected the zero-point energy and these experiments may be referenced through the text by Harris.[7]

INDUCING COHERENCE

How can this energy be tapped? The key is Boyer's observation that matter and the zero-point energy mutually

Table 2

Reasons the Zero-Point Energy Is Difficult to Detect

1. It is incoherent.
2. The energy is everywhere. Its detection requires measuring an energy difference.
3. Less than one quantum of energy is cohered at any one mode.
4. It flows orthogonally to our space (virtual).
5. It rapidly changes frequency. Linear detectors cannot resonate coherently with it to follow signal.
6. The very high frequencies do not readily interact with matter.

interact. This opens the possibility for a positive feedback loop that coheres this energy. Normally, the action of the zero-point energy is random and incoherent. But what type of system can create order from disorder? The thermodynamics advanced by Prigogine[9,10] identifies what types of systems tend toward increasing entropy or randomness and what kinds of systems tend to take random action toward macroscopic order.

Linear systems always tend toward increasing entropy. A linear system is characterized by linear superposition which states that the result of the sum of two inputs is the sum of their corresponding outputs. Since most systems analyzed in science are modeled by sets of linearized equations, it is not surprising that a majority of scientists believe all systems must tend toward increasing randomness and disorder. As a result of this belief a paradox arises. How can one explain the existence of life without violating thermodynamics? The recent work of Prigogine clarifies the second law of thermodynamics. He demonstrates that nonlinear systems under certain conditions may evolve toward macroscopic order. A simple example of this thesis is the rectifier circuit (Figure 1). Here thermal noise from the resistor is channeled through the one-way valve of the diode to charge up the capacitor. Thus, energy in a random state (thermal noise) is channeled to produce energy that can be used for work (charged capacitor).

Since a nonlinear system does not exhibit linear superposition, a combination of inputs often produces surprising,

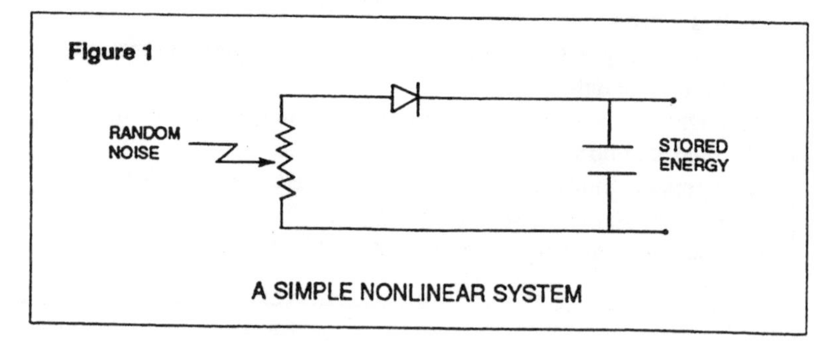

Figure 1

A SIMPLE NONLINEAR SYSTEM

synergistic effects—the whole becomes greater than the sum of its parts. A striking example of this comes from the field of plasma physics. When sufficient energy (e.g., an electric impulse) is added to a gas, it ionizes into a plasma. If more energy is added, the electric charges undergo violent, random, turbulent motion. If still more energy is added, a surprising thing can sometimes occur: The violent turbulent plasma forms up into a meta-stable vortex ring called a plasmoid.[11-13] Figure 2 is a cross-section diagram of the current flow in the plasmoid. Such a structure cannot be predicted by a linear thermodynamic model, but it can be predicted by a nonlinear magnetohydrodynamic model. The nonlinear interactions produce macroscopic coherence from random turbulence.

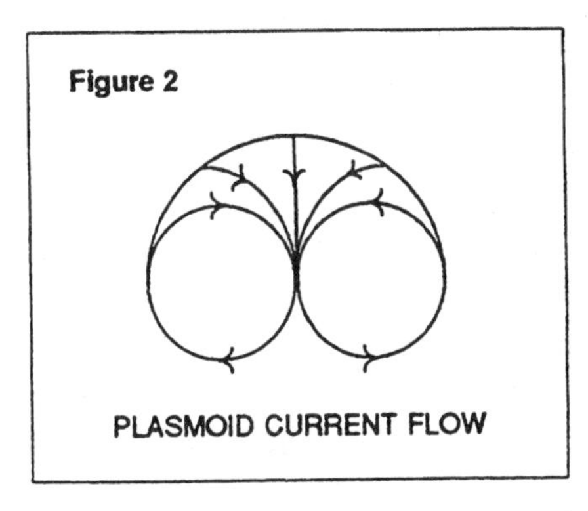

This plasmoid vortex ring may produce a cohering resonance with the zero-point energy as the zero-point energy interacts with the plasmoid. This interaction occurs in a nonlinear system evolving toward meta-stable order. Could the plasmoid slightly cohere the zero-point energy by vacuum polarization so that it provides the energy flux needed to maintain the system? Are there any examples in nature that imply such a thing could occur? Ball lightning has been modeled as a vortex ring plasmoid,[15,16] and its surprising persistence implies it is interacting with some source of energy.

BALL LIGHTNING

Ball lightning appears as a glowing, fireball that sometimes is produced during thunderstorms or in accidents involving electric discharge. The unusual thing about it is its persistence. Most discharges decay very rapidly, but ball

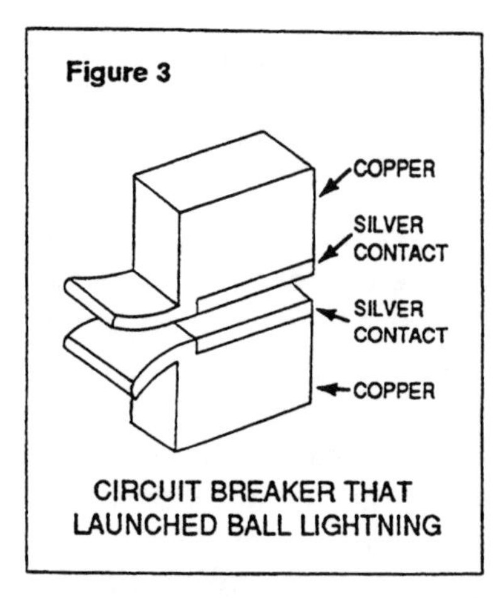

Figure 3

CIRCUIT BREAKER THAT LAUNCHED BALL LIGHTNING

lightning has been observed to last for many seconds[18] Its behavior is unusual too. It sometimes tunnels through windows or travels down chimneys. It has been reported to enter the cockpit window of aircraft, float down the fuselage, and exit through the tail.[18] There is also a case on record where it has been produced by accident more than once in a submarine.[17] A discharge from a specially-shaped circuit breaker (Figure 3) launched a green, glowing fireball. It scared everybody out of the engineer room and then proceeded to float down the corridor before it decayed away. Ball lightning is truly an unusual and surprising phenomenon.

The energy content of this plasmoid has not been adequately explained by conventional physics. Especially difficult to explain is its persistence within a shielded environment (e.g., submarine hull). However, a zero-point energy interaction can explain its persistence, its large energy content, and its surprising tunneling behavior. Moreover, this ball lightning discharge occurs in Gray's[28,30] motor and in some of Moray's[29] corona discharge tubes. It also occurs in the mercury vapor discharge lamps of a local investigator.[31] *All three inventors claim a net energy gain from their devices.*

EXPERIMENT

To verify these claims it would be most useful to repeatably produce ball lightning in the laboratory. Here is suggested an experiment that may do this (Figure 4).

The design was inspired by the inventions of Tesla, Moray and Gray as well as the intimate theoretical connection

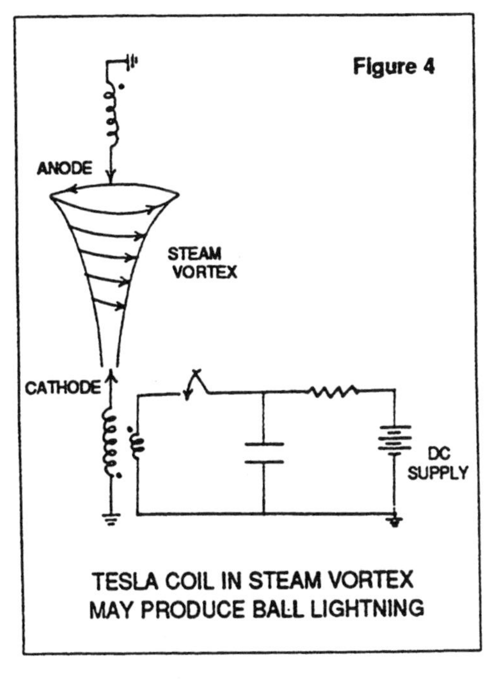

between the soliton and the vortex.[20–22] A soliton is a clumped, nonlinear waveform that tends to maintain its shape. A vortex is like a tornado. Since ball lightning appears to be a soliton form in its ability to maintain itself, why not try to form it from a vortex? Note this experiment resembles the conditions under which a thunderstorm produces ball lightning.

Form a vortex in a rapidly ionizable vapor (e.g., water vapor), then ionize it with an abrupt electric discharge. Here a Tesla coil is suggested to produce the discharge. The anode coil may be wound to produce an opposing magnetic field as in Gray's motor.[30]

The electrode geometry is important. The production of ball lightning can be compared to blowing a soap bubble. It requires precise boundary conditions. Tesla[32] observed ball lightning in his large coils using a spherical electrode. Walters[14] observed a toroidal discharge using a disk cathode. Wells[15] used a cone-shaped plasma gun to produce his plasmoid vortex rings. Silberg[17] wrote an interesting account of the ball lightning accidents on the submarine. The generator circuit breaker had electrodes of a fanning geometry (Figure 3). Here the electrical discharge is forced onto the wide region of the electrodes by a blow-out coil (Figure 5). Both the electrode structure and the pulsed magnetic transient seem significant.

Opposing magnetic fields have been associated with ball lightning production. Tesla launched fireballs from his large coils when the oscillations were phased to create opposing magnetic fields.[44] Both the Gray[30] motor and the circuit breaker blowout coil utilize bucking magnetic fields. Perhaps the most efficient coil structure to create such opposing fields is the "caduceus" wound coil.[45] Here the double helix symmetry of the windings allows for perfect opposition of not only the magnetic fields but their higher order time derivatives as well. Could then opposing magnetic pulses maximize their stress on the fabric of space causing a "hyperspatial involution" that orthorotates the zero-point energy flux? (See appendix.) Abrupt, opposing magnetic transients could be important for ball lightning creation.

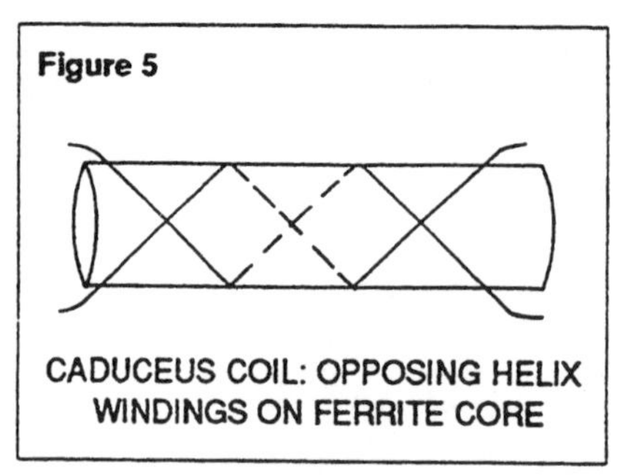
Figure 5

CADUCEUS COIL: OPPOSING HELIX WINDINGS ON FERRITE CORE

The material of the cathode is important as well. Ideally, a large number of electrons should be ejected from the cathode surface simultaneously. Then the deBroglie waves of the ejected electrons could sum constructively to induce a zero-point energy coherence. Moray[29] used an iron sulfide-bismuth junction to produce a luminous discharge. Metallic sulfides are known to form exciton[23] traps; thus, a large number of electrons can be stored in an excited state and then discharged together forming a cohered brush discharge.

Both the cathode geometry, material and magnetic opposition are important for the repeatable production of ball lightning. Another important contributor could be the ionization of the rapidly-moving medium. No one to date has reported ionizing a preformed vortex. The results could be surprising. Once the exact electrode structure is discovered, the creation of ball lightning will become simple and inexpensive.

SUMMARY

Most of the physics literature describe the vacuum as filled with fluctuating energy in some form. I haven't seen any modern literature that claims it is an empty void. Yet most scientists *believe* the vacuum is a void containing no energy whatsoever.

To a physicist who knows about the zero-point energy, the major objection to tapping it is violation of the second law of thermodynamics. However, the recent work of Prigogine has expanded the second law of thermodynamics to include systems that evolve toward increasing order. This, coupled with Boyer's description of the zero-point energy interacting with matter, opens the possibility for a coherence. This may well be experimentally verified when a large number of investigators produce ball lightning thereby causing the recognition of a totally new energy source.

There are other potentially viable methods of tapping the zero-point energy. The concepts of rotation and precession apply directly to elementary particles. They may be considered as "spinor" coherences in the zero-point energy. Simultaneous repetition occurs in the synchronous brush discharge, and soliton-vortex formation applies to the example of ball lightning. The stepdown of the high frequency modes of the zero-point energy shall be the topic of a future seminar. Future work will demonstrate how the plasmoid vortex ring manifests all four of these processes. It is my hope that this discussion inspires research efforts to produce ball lightning, for this may well unlock a new energy source for humanity.

APPENDIX

Conditions for Coherence, Implications of Higher Spatial Dimensions

Prigogine's thermodynamics[9,10] requires two conditions for a nonlinear system to cause random microscopic fluctuations to become cohered macroscopic fluctuations. The first condition requires the system be far away from thermodynamic equilibrium. The second is that the system must be a dissipative structure (i.e., there must be an energy flux through the system in order to maintain it). The key point in question is: Can the zero-point energy provide such a flux in order to maintain ball lightning? This totally depends on the nature of the vacuum fluctuations. It is generally assumed it would require more energy to organize the zero-point energy through vacuum polarization than could be returned by the zero-point energy. Here the zero-point energy is treated as a passive system similar to the polarization of matter. It is clear that no energy could be obtained from such a system.

However, there is evidence that the zero-point energy is not a passive system but actually is a manifestation of an energy flux passing through our space orthogonally from higher dimensions. Wheeler derives such hyperspace channels (wormholes) in his geometrodynamics.[4] Also, a picture of nonlocal connections is implied by quantum physics' EPR paradox,[33,34] Bell's Theorem,[35] and hidden variable concepts.[24] In addition, Sarfatti,[36] Feynman[37] and Dirac[38] describe quantum mechanical propagators summing across the higher dimensions of superspace,[4] a picture that Everett similarly derives in his *Many Worlds Interpretation of Quantum Mechanics*.[26] Note that Everett's theory is derived from a simpler postulate base than standard quantum mechanics (e.g., Von Neumann[39]). Since no special postulates are created to describe the observer, he is treated as a quantum mechanical system like everything else. From this simpler postulate base comes hyperspace, containing an infinite number of three-dimensional universes.

Many physicists have deduced the existence of higher dimensions from independent considerations. An experiment which supports such a concept is the EPR experiment[33,34] which has thus far defied explanations restricted to a three-dimensional universe. There exists considerable discussion in the physics literature implying higher dimensions of space. Moreover, there exists no proof in physics that proves higher spatial dimensions cannot exist.

In general, however, the scientific community has rejected the existence of higher spatial dimensions because human perceptual limitations make it impossible to picture. Note that many quantum mechanical events likewise violate intuition (e.g., tunneling,[40] space-like quantum transitions,[36] two-slit experiment,[41] EPR experiment[33,34]). These can be explained by allowing a greater dimensionality. *A priori* rejection of higher physical dimensions is done by human prejudice, not by scientific proof, and it violates the evidence accumulated by modern physics.

The zero-point energy can be modelled as an electric flux flowing orthogonally through our three-dimensional space (Figure 6). As this flux vibrates, it generates an electric field component in our space creating "mini-white holes" (flux entrances) and "mini-black holes" (flux exits). The random action of this higher dimensional process gives rise to the observed zero-point fluctuations in three-dimensional space. If a plasmoid polarizes the vacuum in a dynamic, nonlinear interaction with the zero-point flux, it could produce a cohered macroscopic fluctuation. This would result in twisting the orthogonal electric flux such that a greater component becomes aligned in our space. Note that quantum theory allows borrowing the energy for a short time period governed by the uncertainty principle: $\Delta E \Delta t \geq h$. This con-

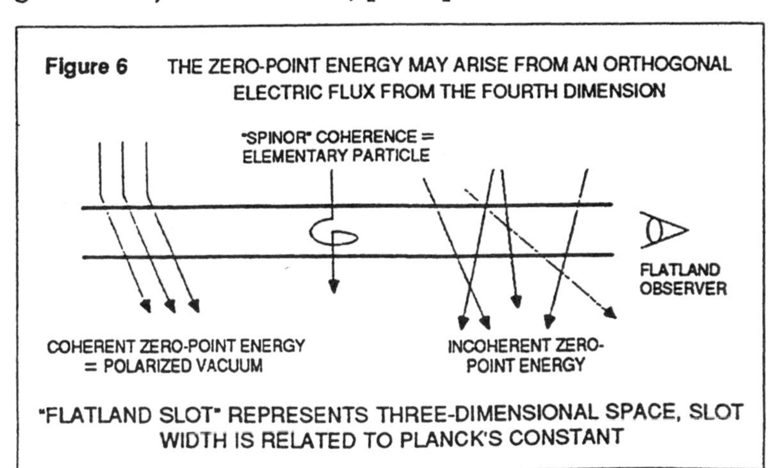

Figure 6 THE ZERO-POINT ENERGY MAY ARISE FROM AN ORTHOGONAL ELECTRIC FLUX FROM THE FOURTH DIMENSION

"FLATLAND SLOT" REPRESENTS THREE-DIMENSIONAL SPACE, SLOT WIDTH IS RELATED TO PLANCK'S CONSTANT

nects the borrowed energy with time. Since general relativity relates the space-time metric to the embedded energy density, could borrowing the zero-point flux locally alter the pace of time?[42] Could the local space-time curvature be altered significantly to produce artificial gravity?[43] These speculations could be experimentally explored by measurements near ball lightning.

The zero-point fluctuations may arise from a higher dimensional flow of electric flux. How else could these fluctuations persist in an expanding universe? Perhaps this process also sustains the elementary particles. Their infinite self-energies appear because the particle has access to this hyperspace energy. The particle is thus a window to the higher dimensional flow. Its finite rest mass results from the amount of energy in our three-dimensional "flatland slot" at any instant. In this view, the elementary particles as well as ball lightning represent resonant modes of the vacuum. The vacuum is not a passive system but a potentially active one. Thus, it can provide the energy flux needed to evolve coherence and maintain ball lightning.

REFERENCES

ETHER, ZERO POINT ENERGY

1. M. Ruderfer, "Neutrino Structure of Ether." *Lett. Il Nuovo Cimento* 13, No. 1, 9 (1975)
   This paper references various Lorentz invariant ether theories.
2. H. C. Dudley, *The Morality of Nuclear Planning*, Kronos Press (1976), Glassboro, NJ 08208.
   This monograph describes a neutrino ether and its relation to radioactivity. Also "Is There an Ether?", *Science Digest*, (May 15, 1975).
3. G. Gamow, *Thirty Years that Shook Physics*, Doubleday, NY (1966).
   This text contains a simple description of Dirac's virtual pair vacuum.

4. C. Misner, K. Thorne, and J. Wheeler, *Gravitation*, W. H. Freeman and Co. (1970).
Chapters 43 and 44 contain description of zero-point fluctuations and superspace. Also J. A. Wheeler, *Geometrodynamics*, Academic Press Inc. (1962) describes vacuum fluctuations and wormholes.

5. T. H. Boyer, "Random Electrodynamics: The Theory of Classical Electrodynamics with Classical Electromagnetic Zero-point Radiation." *Phys. Rev.* D11, No. 4, 790 (1975).

6. M. O. Scully, M. Sargent, "The Concept of the Photon." *Physics Today*, 38, (March 1972)

7. E. G. Harris, *A Pedestrian Approach to Quantum Field Theory*, Wiley Interscience (1972). Chapter 10, "The Problem of Infinites in Quantum Electrodynamics."
This text references experiments that detect the zero-point energy.

8. S. L. Adler, "Some Simple Vacuum PolarizationPhenomenology..." *Phys. Rev.* D10, No. 11 (1974).

NONLINEAR THERMODYNAMICS

9. I. Procaccia, J. Ross, *Science* 198, 716 (18 November 1977).
This article describes Prigogine's Nobel Prize winning work.

10. P. Glandsdorff, I. Prigogine, *Thermodynamic Theory of Structure, Stability, and Fluctuations*, Wiley Interscience, NY (1971).

PLASMOIDS, BALL LIGHTNING

11. *International Journal of Fusion Energy* Vol. 1, No. 1, (1977), No. 3-4 (1978); Fusion Energy Foundation, 231 West 29 St., NY 10001.
Both issues discuss plasmoids and contain abundant references.

12. W. H. Bostick, "Experimental Study of Plasmoids." *Phys. Rev.* 106, No. 3, 404 (1957).

13. D. R. Wells, "Dynamic Stability of Closed Plasma Configurations." *J. Plasma Phys.* Vol. 4, Part 4, 654 (1970).

14. J. P. Walters, *Science* 198, No. 4319, 787 (Nov. 1977).
Walters observes toroidal discharges in his plasma experiments.

15. P.O. Johnson, "Ball Lightning and Self Containing Electromagnetic Fields." *Am. J. Phys.* 33, 119 (1965).

16. M. B. King, "Energy Source Implications of a Helicon Toroid Model for Ball Lightning." *QPR* No. 18, Valley Forge Res. Center, Moore School, University of Pennsylvania (1976)

17. P. A. Silberg, "Ball Lightning and Plasmoids." *J. Geophys. Res.* 67, No. 12 4941 (1962)
This paper describes the circuit breaker that repeatably launched ball lightning in a submarine.

18. S. Singer, *The Nature of Ball Lightning.* Plenum Press, NY (1971).

SOLITONS, VORTICES, EXCITONS

19. A. C. Scott, et al., "The Soliton: A New Concept in Applied Science." *Proc. IEEE,* Vol. 61, No. 10, 1443 (Oct. 1973).

20. S. Bardwell, "The Implications of Non-linearity," Fusion Energy Foundation Newsletter reprint.
This article relates solitons to vortices.

21. G. L. Lamb, "Solitons and the Motion of Helical Curves." *Phys. Rev. Lett.* 37, No. 5, 235 (1976).

22. F. Lund, T. Regge, "Unified Approach to Strings and Vortices with Soliton Solutions." *Phys. Rev.* D14, No. 6, 1524 (1976),

23. R. S. Knox, *Theory of Excitons, Solid State Physics,* suppl. 5, Academic Press, NY (1963).

NONLOCAL CONNECTIONS, HYPERSPACE

24. D. J. Bohm, B. J. Hiley, "On the Intuitive Understanding of Nonlocality as Implied by Quantum Theory." *Found. Phys.* Vol. 5, No. 1, 93 (1975)

25. H. P. Stapp, "Are Superluminal Connections Necessary?" *Il Nuovo Cimento,* Vol. 40B, No. 1 (1977)

26. H. Everett, *The Many Worlds Interpretation of Quantum Mechanics,* Princeton University Press (1973) Also *Rev. Mod. Phys.* 29, No. 3, 454 (1957) "Relative State Formulation of Quantum Mechanics."

27. B. Toben, *Space-Time and Beyond,* E. P. Dutton and Co., NY (1975).
This book is a pictorial introduction to multiply connected space and hyperspace.

INVENTIONS

28. T. Valentine, "Suppressed Inventions," *Newsreal Magazine* No. 2, (1977) National Exchange, P.O. Box 147 Morton Grove, IL 60053.
    This issue contains articles about T. H. Moray and E. V. Gray.
29. T. H. Moray, *The Sea of Energy in Which the Earth Floats*, Cosray Research Institute, 2502 South 4th East St., Salt Lake City, UT 84115. Also U. S. patent 2,460,707 (1949) "Electrotherapeutic Apparatus" contains capacitor corona discharge tubes.
30. E. V. Gray, U.S. Patent 3,890,548 (1976) "Pulsed Capacitor Discharge Electric Engine."
31. G. Obolensky, Private communication (1977).
32. N. Tesla, *Lectures, Patents and Articles*, Nikola Tesla Museum, Beograd (1956).

QUANTUM MECHANICS

33. A. Einstein, B. Podolsky, N. Rosen; "Can Quantum Mechanical Description of Physical Reality be Considered Complete?" *Phys. Rev.* 47, 777 (1935).
34. S. J. Freedman, O. F. Clauser, "Experimental Test of Local Hidden Variable Theories." *Phys. Rev. Lett.* 28, 938 (1972).
35. H. P. Stapp, "Bell's Theorem and World Process." *Il Nuovo Cimento*, Vol. 29B, No. 2, 270 (1975).
36. J. Sarfatti, "Implications of Meta-Physics for Psychoenergetic Systems." *Psychoenergetic Systems*, Vol. 1, 3 (1974).
37. R. P. Feynman, "Space-Time Approach to Quantum Electrodynamics." *Phys. Rev.* 76, 769 (1949).
38. P. A. M. Dirac, "The Lagranian in Quantum Mechanics." Reprinted in *Quantum Electrodynamics*, ed. J. Schwinger, Dover Publications, NY (1958).
39. J. Von Neumann, *Mathematical Foundation of Quantum Mechanics*, Princeton University Press (1955).
40. B. Josephson, "The Discovery of Tunneling Supercurrents." *Science* 184, 527 (May 3, 1974).
41. R. P. Feynman, A. R. Hibbs, *Quantum Mechanics and Path Integrals*, McGraw Hill, Inc. (1965).

42. N. A. Kozyrev, "Possibility of Experimental Study of the Properties of Time," Sept. 1967. *JPRS* 45238, U.S. Dept. of Commerce, National Technical Information Service, Springfield, VA 22151.

43. M. B. King, "Is Artificial Gravity Possible?" (May 1976); "Is Antigravity Possible?" (Dec. 1975). Moore School of Electrical Engineering, Dept. of Systems Engineering, Univ. of Pennsylvania, Phila., PA 19104.

44. H.W. Secor, "The Tesla High Frequency Oscillator." *Electrical Experimenter,* 3, 615 (1916).

45. J. Bigelow, D. Reed, Private Communication (1977).

46. M. Ruderfer, "Comments on A New Experimental Test of Special Relativity..." *Lett. Nuovo Cimento* 3, 658 (1970).

IS ARTIFICIAL GRAVITY POSSIBLE?

May 1976

ABSTRACT

Inducing a slight coherence in the action of the zero-point energy may curve the space-time metric yielding artificial gravity. The unidirectional thrust exhibited by stressed, charged dielectrics in the experiments of T. Townsend Brown may be evidence for this. A plasma vortex might enhance this effect for practical applications.

Is artificial gravity possible? If so, it would be an attractive means of propulsion because it would produce rapid acceleration without stress. According to general relativity, energy curves space-time producing gravity. If sufficient energy is placed above you, it can make you fall up. The mass equivalent of energy needed for levitation is about 10^{12} grams. If we have to provide all this energy, artificial gravity would be beyond today's technology.

However, modern quantum physics has within it an astonishing construction. It is the existence of the zero-point vacuum energy. Empty space is not empty. It consists of fluctuations of electricity whose energy density is on the order of 10^{94}grams/cm^3—an enormous number. This energy normally is unobserved because it self-cancels by destructive interference. However, if a device could induce just a slight coherence in the action of this energy over a region of space, then it could produce artificial gravity.

The work of T. Townsend Brown may give us a clue to how this might be done. A sufficiently charged capacitor can cause a vacuum polarization—a slight coherence of the vacuum fluctuations. Also the ionic lattice of a rapidly spinning body may interact with the vacuum energy causing a slight coherence which would alter the inertial properties of the body. This would occur because the vacuum energy itself curves space-time. Figure 1 illustrates the curvature of space-time with a two dimensional plane representing three-dimensional space. The lines are the path light takes as it travels through space. A massive body warps space, causing the path of light to bend. Energy also causes space to curve, as Einstein's theory of general relativity describes.

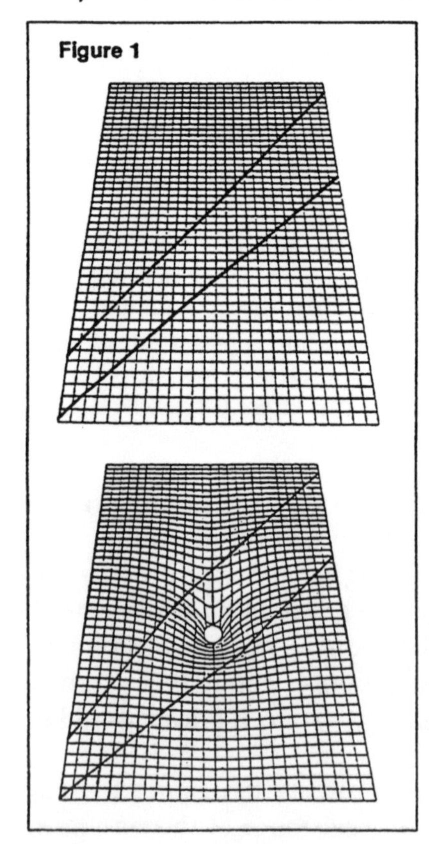
Figure 1

In Figure 2 the block diagram represents the ten nonlinear differential equations of general relativity. The box *T* represents the stress-energy tensor. It describes the location and flow of energy in space-time. The box *g* represents the metric. It de-

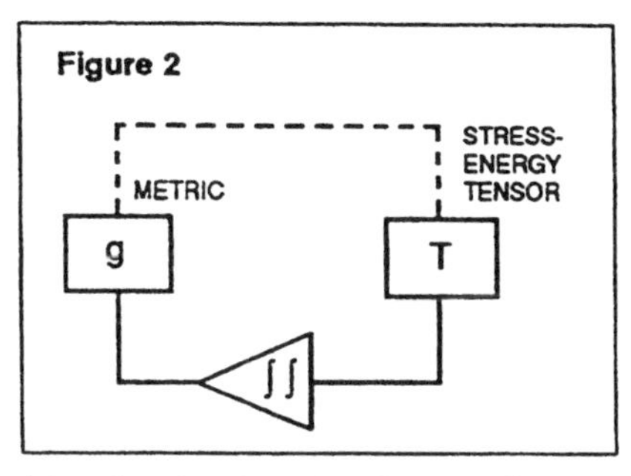

scribes the amount of space-time curvature that the stress-energy tensor induces. The double integration indicates that the stress-energy tensor controls second derivative terms of the metric. The dotted line represents an idea by Andrie Sakharov: The metric elasticity of space governs the action of the zero-point vacuum fluctuations, an energy which must be included in the stress-energy tensor. *This creates a feedback loop on a potentially active system.*

Can this system resonate? The nonlinearities of the system imply this may be possible. To illustrate, consider two boxes of energy and the curvature they induce at the point *P* (Figure 3). If we move the energy box *A* while holding the box *B* stationary, box B's contribution to the curvature at point *P* varies. If the system were linear, superposition would apply and B's contribution to the metric would be independent of A's. However, the system is nonlinear. For certain locations of A, B's contribution to the curvature is maximized. This idea also applies to a continuous field. As the field shape varies, the amount of curvature varies due to the mutual self-interaction of the field components. A resonant field is that field shape which maximizes the curvature of space-time. To efficiently curve space-time it is not only the amount of energy that is important, but *how this energy is used.* What is the coupling mechanism that allows mutual interaction of distant energy increments? It clearly has to be the fabric of space itself.

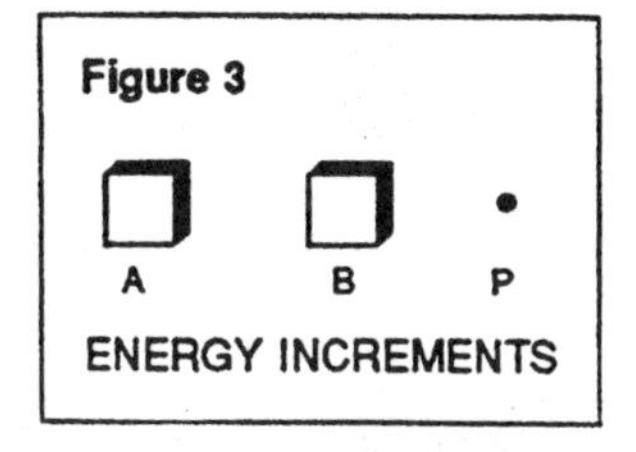

Microscopically viewed, space is a turbulent sea of energy consisting of electric flux (Figure 4). This flux enters from higher dimensional space through "mini-white holes" and leaves our three-dimensional space through "mini-black holes." To picture this concept, imagine our existence is confined to the two dimensional planar universe, flatland. We have no awareness of a third dimension. If a flux of energy were to pass through our space perpendicularly, we would have no awareness of this energy. However, if this flux jitters as it passes through flatland, a component of its motion would exist in our space. This flux component is the zero-point vacuum fluctuations. The diameter of these "mini holes" is on the order of Planck's length, 10^{-33}cm. The energy density through them is enormous, 10^{94}grams/cm^{3}.

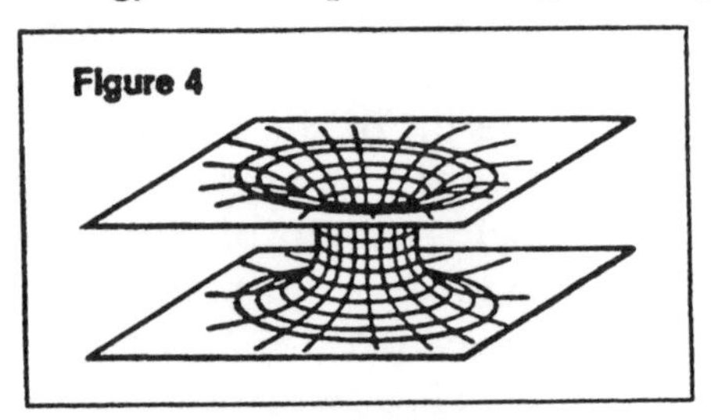
Figure 4

Large energy densities cause gravitational collapse. The top plane of Figure 4 represents three-dimensional space. The bottom plane also represents three-dimensional space —perhaps the same space. A large energy density causes space to pinch into what John Wheeler calls a *wormhole*. A wormhole can channel electric flux through higher dimensional space. It can connect distant points in the same three-dimensional space. In Figure 4, the plane represents three-dimensional space; the tube is the wormhole. Electric flux entering results in a "mini-white hole", flux exiting is a "mini-black hole." "Mini-holes" are constantly being created and annihilated in space, causing changing wormhole connections. Wheeler calls this resulting multiply-connected space *superspace*.

Can the vacuum fluctuations cohere in a region of space? Unstable particles or resonances may result from the temporary local coherence of the vacuum fluctuations. This model implies the existence of a whole spectrum of very small subnuclear particles that our science has not yet detected. The charge of a particle depends on a predominant flux from one type of mini hole.

The stable particles may likewise be a coherent alignment

of these holes. This model begets an interesting interpretation of the electron cloud around the nucleus of an atom. The electron literally is a cloud of negatively biased vacuum energy that maintains itself by a cohering self-connectivity through wormholes. This interpretation may shed insight on the wave-particle duality of matter. It illustrates coherence of the vacuum fluctuations in the quantum world.

Can the vacuum energy cohere over a large region of space in the macroscopic world? Note that levitation needs only a slight coherence in a statistical sense since the 10^{12} grams required for levitation is so much smaller than the 10^{94}g/cm^{3}. How can we achieve macroscopic vacuum polarization? Perhaps the work of T. Townsend Brown gives a hint.

Basically, Brown discovered that a sufficiently charged capacitor exhibits unidirectional thrust in the direction of the positive plate, and some types of capacitors exhibit more thrust than others. A type that worked very well consisted of 10,000 layers of lead foil and insulator. A dielectric consisting of a mixture of lead oxide and resin also worked well. Experiments with other materials lead Brown to the conclusion that a more massive dielectric with a greater dielectric constant produces a greater thrust.

T. Townsend Brown realized that the air around the capacitor's positive plate could be ionized and the fringe field would accelerate these ions back toward the negative plate causing the capacitor to move. In fact, J. Frank King, a collegue of Brown, patented a vehicle propelled by this type of ion propulsion (Figure 5). The top ring (21) ejects a plasma, and the rings (14, 15, 16) produce a magnetic field synchronously timed to accelerate the plasma downward.

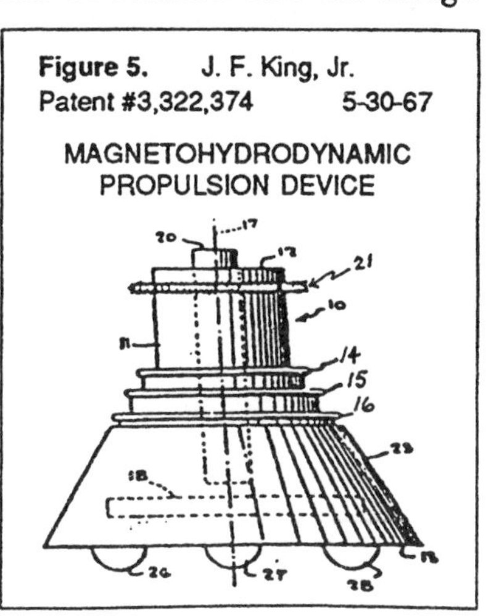

Figure 5. J. F. King, Jr.
Patent #3,322,374 5-30-67

MAGNETOHYDRODYNAMIC PROPULSION DEVICE

The reaction force accelerates the vehicle upward.

To demonstrate that his capacitors involve something more than ion propulsion, T. Townsend Brown immersed them in oil, a medium that does not readily ionize. He observed that the thrust was almost the same as in air indicating that ion propulsion was not the major component of thrust. Brown charged the oil tank to the same potential as the positive plate in order to rule out electrostatic attraction as the cause of the thrust.

T. Townsend Brown also tested capacitors in a vacuum. He mounted two aluminum, open gap, parallel plate capacitors on a rotor. The vacuum pressure was monitored and held steady at 10^{-6} Torr. As he gradually increased the voltage from 90KV to 200KV, he observed irregular sparking concurrent with a large thrust. He also observed a residual thrust in the absence of sparking. The sparking occurred initially at about 15 second intervals. Its frequency gradually decayed until after about five minutes of operation, no more sparking occurred even though he left the rotor running days at a time. At 200KV the angular velocity would continue to increase, and he had to reduce the voltage to prevent the rotor from flying apart.

Running the rotor for days at a time that lead T. Townsend Brown to a remarkable observation. The capacitor thrust varied with the time of day even though the voltage, temperature, and pressure were held constant and carefully monitored. Over weeks of operation he observed a distinct sidereal correlation in the amount of thrust. This led Brown to believe that the charge capacitors were like catalysts that caused a vacuum polarization interaction with some type of energy flux hitting the earth from space. Perhaps the energy is from the sun; perhaps it is from the center of the galaxy. Brown is currently working with the Stanford Research Institute to determine the nature and source of this energy.

During the early 1940's T. Townsend Brown made a curious discovery. He found that enlarging and curving the positive electrode increased the thrust, and later he patented this concept (Figure 6). In the patent the large positive electrode is labeled (12); the negative electrode (14) and the dielectric rod connecting them (10). During World War II Brown discovered the optimum electrode shape. He described it as *triarcuate*—meaning "three arcs." He used a system of weights and pulleys to measure the thrust (Figure 7). When charged, a bright, colorful corona would appear on the surface of the triarcuate aluminum canopy.

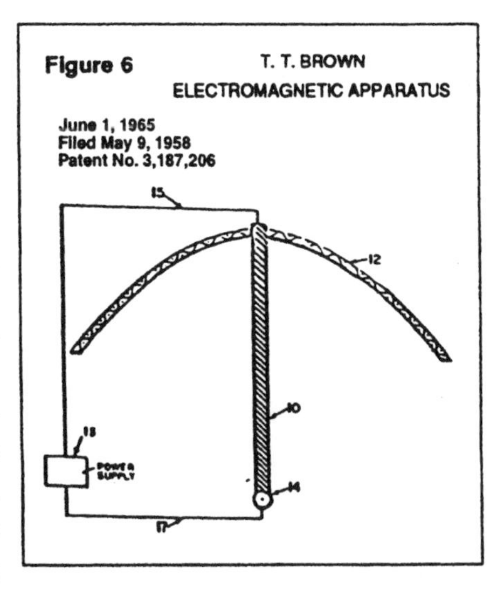

The factors that increased the thrust on the capacitor in T. Townsend Brown's experiments are:

1. Increase the plate area
2. Decrease the distance between the plates
3. Increase the dielectric permittivity
4. Increase the voltage
5. Increase the mass of the dielectric
6. Shape the positive plate

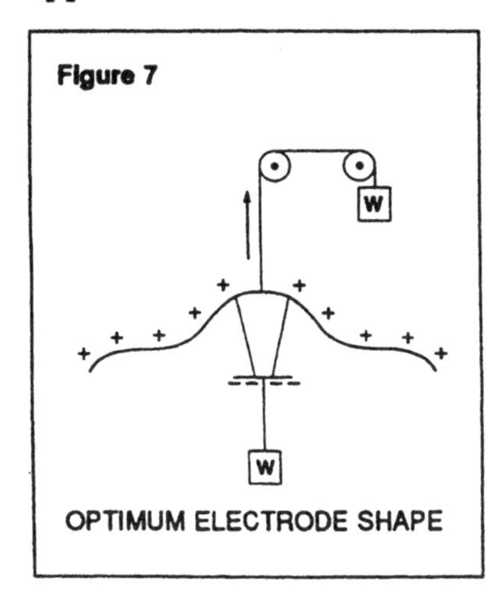

The first three factors increase the electrical capacitance of the apparatus. The thrust was approximately linear in voltage over the tested range of 50-300KV. Point 5 is what Brown feels links the nature of the capacitor thrust to gravity. Point 6 needs to be explained.

Any hypothesis that explains the thrust must include the

following key observations:

1. A large thrust was associated with a spark. A residual thrust existed without sparking. (In vacuum, 1956, gap capacitors).
2. A DC voltage (150KV) caused a thrust when initially applied. The thrust would decay within 60 seconds. Two minutes of recovery were needed at zero volts before thrust could be produced again. (In oil, 1928, dielectric of lead oxide and wax).
3. The thrust varied with the time of day. (In vacuum and in oil).

Point 1 requires identifying the source of the vacuum spark. Could it be due to air molecules that were trapped in the positive plate, due to electrons ejected from the negative plate, or due to both?

The second point was only observed in Brown's early capacitors of lead oxide and wax. It provides a clue to the optimum operating voltage for the capacitors. The dielectric should be polarized to the threshold of breakdown such that perturbations can cause avalanche breakdown. If the voltage is too high and the dielectric conducts, there will be no thrust. The thrust is associated with a change in state from polarization to breakdown. If the voltage is adjusted such that this change of state keeps repeating, a maximum thrust occurs.

Point 3 is a surprise and needs explanation.

Some possible hypotheses to explain the observations are listed below:

1. The surrounding medium is ionized and accelerated by the field. (Ion propulsion).
2. Avalanche breakdown through the dielectric is associated with:
   a. Plasma formation in the dielectric.
   b. An abrupt change in polarization.
   c. An abrupt change in the dielectric permittivity. A rapidly modulated dielectric permittivity may act as a transducer between electromagnetism and any

of the following:
1. Zero-point vacuum energy.
2. High frequency gravitational radiation.
3. High frequency permittivity waves.
4. Higher dimensional components of electromagnetism.
5. Neutrino flux
6. Ether flux

3. A resonant field is produced. The positive electrode is shaped to maximize the mutual interaction of the field with the metric. This may result in a spatially extended coherence of the vacuum energy fluctuations yielding a macroscopic metric fluctuation.

Ion propulsion is a component of thrust, but it cannot explain all the behavior. To illustrate, consider two equal size capacitors, the first with a small dielectric constant and the second with a massive material that has a large dielectric constant. Apply the same voltage to both. Brown observed that the capacitor with the larger dielectric constant exhibits the greater thrust. This is exactly opposite to what ion propulsion would predict since the fringe field of the first capacitor is greater. In the vacuum rotor experiment, only residual air ions accelerated in the fringe field can cause ion propulsion. The air ions in the main field will bang into the negative plate and impede the thrust. However, air ions in the main field between the plates can trigger breakdown—causing an electron cloud ejection from the negative plate.

The key to artificial gravity may be a rapidly accelerated, densely charged plasma cloud. It might cohere the vacuum fluctuations over a macroscopic region of space if there exists mutual coupling and connectivity of the particles in the cloud. Wheeler's superspace shows how a nonlocal connectivity may occur.

A simultaneous avalanche breakdown across the entire dielectric might macroscopically cohere the vacuum fluctuations in the region. If this were to occur in a perfect crystalline substance, the coherence could be significantly larger due to the regular ionic lattice's coupling to the vacuum energy. An outside energy flux could trigger and coherently couple the particles participating in the simultaneous breakdown.

What possible energy source could be interacting with the polarized field to account for the sidereal correlations? Could it be high frequency gravitational waves? Or could there exist such a thing as permittivity waves, perhaps generated by a plasma modulating a medium's dielectric constant? Brown ruled out standard electromagnetism by recent shielding experiments. But could there exist a higher dimensional form of electromagnetism that penetrates shielding? Could neutrinos be interacting? Could an ether flux exist that Michelson and Morley did not detect because the flux was perpendicular to the plane of their interferometer? (See references 5, 16, and 17.) These ideas are obviously speculative—only future experiments can give clues to develop a more concrete formulation.

Experiments by Bruce DePalma, N. A. Kozyrev and W. J. Hooper may provide a hint on how to magnify the effect. *Spin is the key*. Rapid spin of a plasma cloud or plasma toroid may result in a dynamic circular vacuum polarization. In a longitudinal magnetic field a plasma will naturally take the form of a spiral—a macroscopic helicon cloud. The optimum electrode structure can shape a helicon plasma into a vortex while guiding it into the fringe field. A plasma vortex may produce a macroscopic resonant field that slightly coheres the zero-point vacuum fluctuations to produce artificial gravity. (Figure 8) Also an inner, solid state plasma helicon through a dielectric or semiconductor may produce a strong vacuum energy coherence. The two plasma spirals may produce

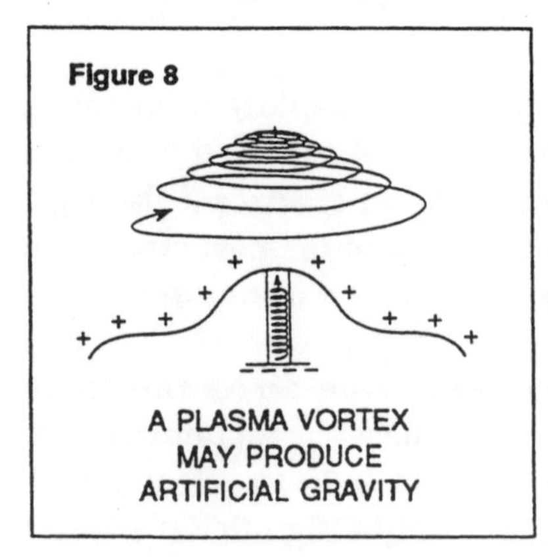
Figure 8

A PLASMA VORTEX MAY PRODUCE ARTIFICIAL GRAVITY

a toroidal vacuum polarization which may affect the inertia of neutral bodies in the region within. A pulsed ion vortex may produce artificial gravity sufficient for practical applications.

It is my hope that this discussion motivates others to continue the experimental investigations started by T. Townsend Brown, for a new propulsion technology awaits our discovery.

APPENDIX

On Coherence of the Vacuum Fluctuations
A Postulate of Physics

Artificial gravity relies on the coherence of the zero-point vacuum energy fluctuations. Many physicists believe that it is impossible to cohere the vacuum fluctuations because this would be a violation of the law of entropy. The law of entropy applies to those systems whose behavior is governed by a large set of *independently* acting components. Entropy is a statistical law stating that the chance alignment of random, independently acting elements is small. The probability approaches zero as the number of independent elements increases.

What is the true nature of the vacuum fluctuations? Are they independent "blinkers," or can an underlying connectivity occur, as described by Wheeler's superspace? The answer to this question is crucial in order to properly apply the law of entropy. In the case of random independent "blinkers," the law applies. The law of entropy may also apply to most cases in Wheeler's superspace formulation as long as the connectivity is random and nonlocal. However, if a device were to bias the connectivity in a region of space, the underlying assumption of *independence* would no longer apply, and it would be improper to invoke the law of entropy. *There is no proof in physics that states such connectivity is impossible.* On the contrary, there exist experiments that imply a vacuum energy coherence is occurring. (See references 7, 8, 9, 10, and 15.)

What is the true nature of the vacuum fluctuations? A postulate—that the vacuum fluctuations are random and independent—has divided modern physics into two camps. Most physicists today believe this postulate, and that it is improper to ask questions about the underlying causality. On the other hand, David Bohm, Jack Sarfatti and Fred Wolf postulate the existence of a possible underlying connectivity (as illustrated, for example, by Wheeler's geometrodynamics). This idea has been developing over the past twenty years and many physicists are unaware of it and its implications. The concepts are somewhat alien to classical physics, and difficult for many to understand since they invoke the existence of a physically real, higher dimensional space. For this reason, this postulate is currently not as popular.

But what governs physics—popularity or experimentation? There does exist a number of experimental anomalies[7–10,15] that can be explained by invoking the latter postulate that can't be explained in other ways. Most physicists have ignored these experiments, but this should not prevent capable individuals from repeating and verifying the work.

It is my hope that scientists will keep an open mind toward these investigations since recent theoretical developments in physics allow the possibility for an experimental success that could yield a tremendous technological advancement for mankind.

REFERENCES

1. Misner, Thorne, and Wheeler, *Gravitation*, W. H. Freeman, NY, 1970.
   An excellent description of the zero-point vacuum energy fluctuations given in Chapter 43 and 44.
2. J. A. Wheeler, *Geometrodynamics*, Academic Press, Inc., 1962.
   Geons and the vacuum fluctuations are described.
3. Toben, Sarfatti, and Wolf, *Space-Time and Beyond*, E. P. Dutton and Co., 1975.
   A layman's introduction to multiply-connective space-time and vacuum energy is illustrated with numerous drawings.

4. R. Wald, "Gravitational Spin Interaction," *Physical Review* D, Vol. 6, No. 2, July 1972, p. 406.
The spinning body gravitational interaction is analyzed.

5. H. C. Dudley, "Is There an Ether?", *Industrial Research,* November 15, 1974; also in *Science Digest,* May 15, 1975, p. 57.
A neutrino flux model for the ether is presented.
See also *Il Nuovo Cimento,* Vol. 4B, No. 1, 68, (1971).

6. P. Bandyopadhyay and P. R. Chauduri, *Nuovo Cimento,* 38, 1912 (1965); 66A, 238 (1969).
Weak coupling of the neutrino and the photon is discussed.

7. C. F. Brush, *Am. Phil. Soc.* V. 67, 105, (1928).
This paper describes an experiment which shows that aluminum silicate falls slower than other materials.

8. W. J. Hooper, *New Horizons in Electric, Magnetic and Gravitational Field Theory.* Electrodynamic Gravity, Inc., 543 Broad Boulevard, Cuyahoga Falls, OH 44221.
The author relates the motional electric field to gravity.

9. N. A. Kozyrev, "Possibility of Experimental Study of the Properties of Time," Sept. 1967, *JPRS* 45238, U.S. Department of Commerce, National Technical Information Service, Springfield, Virginia 22151.
An experiment relating the spin of mass to the pace of time is discussed.

10. B. E. DePalma, "A Simple Experimental Test for the Inertial Field of a Rotating Mechanical Object," *Journal of the British-American Scientific Research Association,* Vol. VI, No. II, June 1976.
The experiments are also described in the appendix of R. L. Dione, *Is God Supernatural?,* Bantam, NY, 1976.

11. C. C. Chiang, "On a Possible Repulsive Interaction in Universal Gravitation." *The Astrophysical Journal,* 1985, 87 (1973).

12. H. Bondi, "Negative Mass in General Relativity," *Rev. Mod. Phys.* 29, No. 3, 423, (1957).

13. J. A. Wheeler, "On the Nature of Quantum Geometrodynamics," *Ann Phys. 2,* 604, (1957).

14. E. Streerwitz, *Phy. Rev.* D 11, No. 12, 3378, (1975).
The author's analysis includes vacuum fluctuations in the stress-energy tensor.

15. S. L. Adler, "Some Simple Vacuum Polarization Phenomenology..." *Phy. Rev.* D 10, No. 11, 3714, (1974).
16. J. Schwinger, "On Gauge Invariance and Vacuum Polarization," *Physical Review* 82, No. 5, 664, (1951).
17. Brill and Wheeler, "Interaction of Neutrinos in Gravitational Fields," *Rev. Mod. Phy.* 29, 465, (1957).
This paper describes the interaction of neutrinos and gravitons.
18. P. A. Dirac, *Roy. Soc. Proc.* 126, 360 (1930).
This paper first introduced the vacuum energy as an electron sea.
19. Gamow, *Thirty Years that Shook Physics*, Doubleday, NY, 1966.
This text contains a layman's description of Dirac's Theory.
20. M. F. Hoyaux, *Solid State Plasmas*, (1970).
This monograph is a concise introduction to solid state physics and plasma physics.
21. D. Bohm, "A Suggested Interpretation of the Quantum Theory in Terms of Hidden Variables," *Phys. Rev.* 85, 166, 180 (1952).
22. L. deBroglie, "The Reinterpretation of Wave Mechanics," *Foundation of Physics* 1, 1-5, (1970).
23. L. Motz, "Cosmology and the Structure of Elementary Particles," *Advances in the Astronautical Sciences*, V8, (1962).
24. H. Stapp, "S-Matrix Interpretation of Quantum Theory," *Phys. Rev.* D3, 1303, (1971).
The S-Matrix "web" describes connectivity.
25. Hawkins and Ellis, *The Large Scale Structure of Space-Time*, Cambridge University Press, 1973.
The basic physics of black holes is discussed.
28. D. Sciama, "Gravitational Waves and Mach's Principle," Preprint IC/73/94 from the *International Center for Theoretical Physics*, Trieste, Italy, 1974.
27. "Physics Made Simple," *Science News*, 106, 20 (July 1974).
This article mentions experimental evidence that shows elementary particles are mini black holes.

STEPPING DOWN HIGH FREQUENCY ENERGY

December 1981

Most oscillating systems have energy associated with their frequency as well as their amplitude. Though energy of frequency does not apply to resistive electrical circuits, it occurs in nonlinear reactive circuits. The appropriate nonlinear reactive circuit can irreversibly and coherently shift high frequency energy down the spectrum with a gain in amplitude. The key reactive components in such a circuit are plasma tubes tuned to resonate at the heavy ion frequency. This system may tap the zero-point energy as a source.

INTRODUCTION

Throughout nature there exist oscillating systems whose energy content is associated with the oscillation's amplitude and frequency. However, in standard linear electrical engineering only the energy associated with a signal's amplitude is recognized. This is correct when describing resistive loads. An exception arises when considering purely reactive elements in low-loss nonlinear circuits. Here energy content associated with frequency is recognized, and if the correct type of system is created, this energy can be synchronously stepped down the spectrum with a gain in amplitude. Further, the underlying principle governing how to build such systems has already been established in the scientific literature. Here it is proposed that if a purely reactive nonlinear system absorbs energy from its environment by irreversibly shifting its frequency down the spectrum, the action of the zero-point energy may resupply the depleted modes.

ENERGY OF FREQUENCY

Most oscillating systems have energy associated with their frequency as well as their amplitude. For example, the simple harmonic oscillator (figure 1) has a resonant frequency at $\omega = \sqrt{k/m}$. The energy of the system is $\frac{1}{2}kX^2$ where X is the amplitude of the oscillation. For the same amplitude, the greater the value of k, the higher the resonant frequency and the more energy stored.

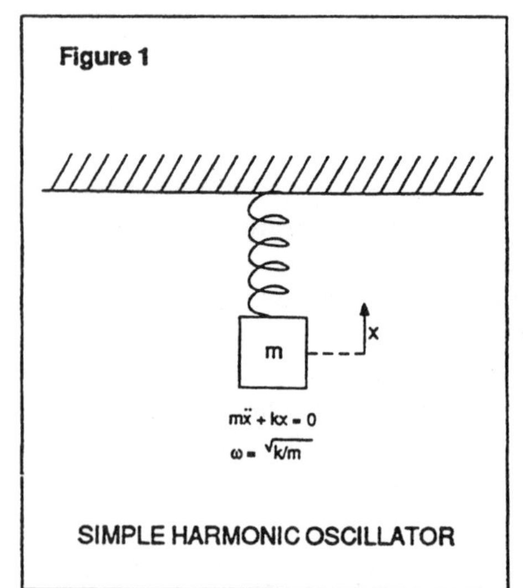

Another example is the description for power delivered by waves on a string[1]. It is proportional to the square of the amplitude as well as the frequency. Also, Rebbi[2] describes the energy of solitons and relates it to their amplitude, velocity and frequency. Of course, the most obvious description of energy related to frequency occurs in quantum mechanics: $E = h\nu$. In principle, this energy could be converted to energy of amplitude by the following idealized system (figure 2).

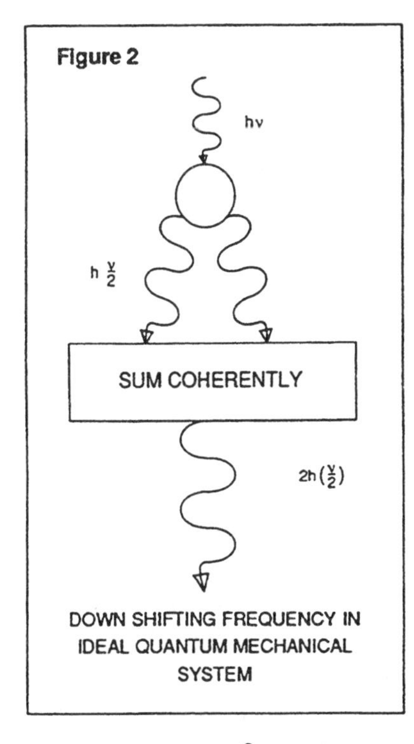

A photon of energy *hv* is absorbed in a quantum mechanical system which reemits two photons at half the frequency. In principle, these two photons could be captured and summed coherently. Thus energy stored in frequency can be converted to energy of amplitude.

Energy of frequency is generally not used by electrical engineers in describing circuits. Here the power associated with an electrical signal is:

$$P = \frac{1}{T}\int vidt = \frac{1}{T}\int \frac{v^2}{R}$$

This equals V_{rms}^2/R for resistive loads. The power is related to the amplitude of the driving voltage, not the frequency. This is due to the losses in resistive loads. In a wire during any half cycle of applied voltage, the electrons undergo many collisions; so many, in fact, that the macroscopic description of the electron's response to an impressed voltage is not acceleration but rather an extremely small drift velocity. The electron never "senses" the frequency of the driving voltage. There is no absorption or storage of energy associated with frequency. In fact, if anything, a typical conductor transduces the impressed power to a very high-frequency incoherent form: infrared radiation (heat). An electrical conductor does exactly the opposite of stepping down energy of frequency; it steps it up. Resistive elements could never sense—let alone step down—energy associated with frequency.

There is an exception where energy of frequency is

recognized in certain types of ideal electrical circuits. In 1956, Manley and Rowe[3] published a set of equations relating power and frequency in various lossless oscillating modes interacting through a nonlinear reactance. Weiss[4] and later Brown[5] derived these same relations from quantum mechanical considerations. Sturrock,[6] Penfield,[7] and Scott[8] have also recognized these relations. If it is assembled correctly, an electrical system involving pure reactance (i.e., inductance or capacitance with minimal resistance) that couples its oscillating modes through low-loss nonlinear elements can coherently convert energy of frequency to energy of amplitude.

THE "SPECTRAL DIODE"

Such a system is described by Nayfeh and Mook.[9] They analyzed a set of differential equations that characterize a system that channels energy from high-frequency modes to low-frequency modes irreversibly, i.e., a kind of "spectral diode" or "frequency domain diode." In this process, there is amplitude gain: "...the second mode is excited, the amplitude of the fundamental mode can be five times the amplitude of the excited mode."[9] The system is described by a set of cubic nonlinear differential equations tuned to a condition of internal resonance:

$$\ddot{U}_1 + \omega_1^2 U_1 = U_1^3 + U_1^2 U_2 + U_2^2 U_1 + U_2^3$$

$$\ddot{U}_2 + \omega_2^2 U_2 = U_1^3 + U_2^2 U_1 + U_1^2 U_2 + U_2^3 + \cos \Omega t$$

where

U_1 = function describing fundamental mode
U_2 = function describing second mode
ω_1 = frequency of fundamental mode
ω_2 = frequency of second mode
$\cos \Omega t$ = driving function
$\ddot{U}$ = second time derivative of $U(t)$

The internal resonance condition is $\omega_2 = 3\omega_1$.

Note the right and left sides of the equations describe a harmonic oscillator, while the right side contains terms of cubic order. Note also the absence of the linear (damping, resistive) term. Although only two stages were analyzed in the text, many stages could be cascaded together to transduce very high-frequency energy down the spectrum with amplitude gain.

A crude circuit embodiment of a similar system would be lossless oscillators with adjacent stages connected by a switch (figure 3).

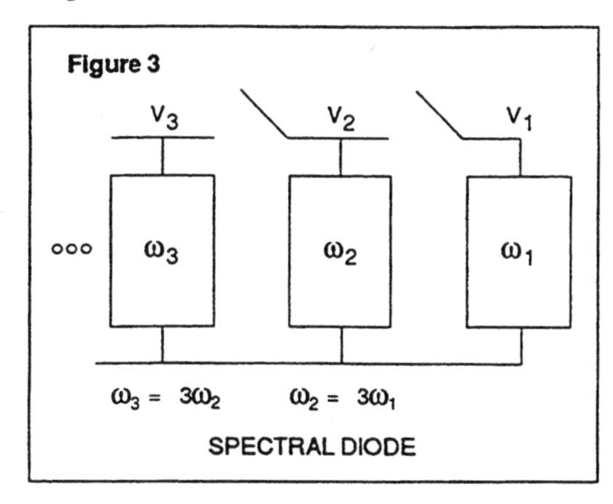

The harmonic oscillators are tuned to a condition of internal resonance ($\omega_{i+1} = 3\omega_i$). The switch is closed for an instant only when the relative phases are such that the current is pulsed from the high-frequency stage to the low-frequency in phase with the low-frequency oscillations. This will synchronously augment the amplitude of the low-frequency stage. The switching is timed so that the energy only flows in one direction. The switch is closed when

$$|V_{i+1}| > |V_i| \text{ and } V_{i+1} V_i > 0$$

This occurs during the shaded regions of figure 4. In a two-stage system the maximum amplitude gain is three to one. However, many stages can be cascaded together to improve this.

Figure 4 shows the advantage of the three to one frequency ratio between the stages. It provides the most pulsing opportunities to excite the lower frequency stages per cycle.

Each stage is excited across the node of its oscillation and it is excited by all their higher frequency stages simultaneously. This system has a tremendous advantage over rectifying noise directly onto a capacitor, for the driving pulse's amplitude can be much less than the oscillation's amplitude. Also the three to one amplitude saturation ratio between stages can be exceeded if a pulse sharpening network is placed between the stages. The saturation point is determined by the rise time of the pulse. This system is designed to drive the high frequency energy down the spectrum as rapidly as possible in order to optimally absorb the available energy.

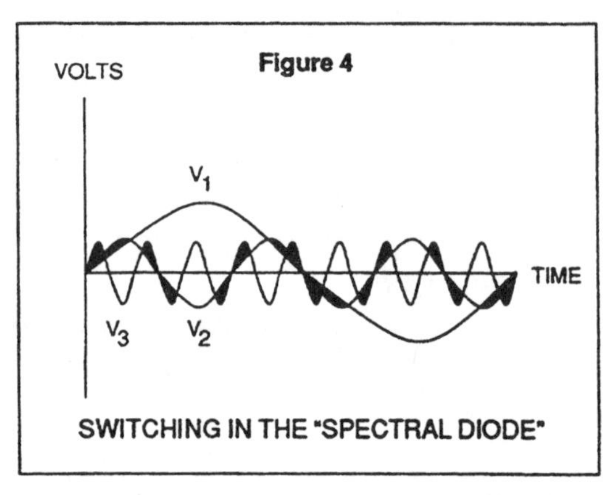

This idealized system required lossless oscillators. It can not be efficiently realized with today's off-the-shelf capacitors and inductors. However, an active medium can be created that will fulfill the conditions of low-loss resonance so that the desired energy transduction will occur.

PLASMA TUBES

To efficiently store and shift high frequency energy down the spectrum, the charge carriers must oscillate at the desired frequency with minimal disruption by collisions. In this way weak pulses injected in phase with the oscillation will be absorbed into the resonating system and tend to augment its amplitude. If collision losses are greater than the injected energy, that energy would be lost. Electrons make poor carriers for this purpose since they are so mobile that their displacement is too large, and they undergo too many collisions. A better carrier would be a proton, or better still, a heavy ion. Because of the heavy ion's mass, the oscillation's displacement would be small; yet the energy stored would still be considerable, for the kinetic energy ($\frac{1}{2}mv^2$) and momentum would be associated with the large mass, not the velocity. As a further advantage, there would be little disruption to the oscillations due to collisions with electrons. The heavy ion velocity would hardly change during such collisions because their mass is so great. Instead such collisions would tend to ionize the gas and maintain the plasma. Thus the electron collisions would be helpful as long as they were not too energetic. Such a system is a plasma tube tuned to the resonant frequency of the (positive) heavy ions. Moray[10] may have been the first to recognize the advantage of this to achieve an efficient step down in frequency.

Normally a plasma tube resonates at its electron plasma frequency, since the electrons are the most mobile carriers. The plasma frequency is approximately .

$$\omega_p = \sqrt{\frac{NQ^2}{\epsilon_0 m}}$$

where ω_p = plasma frequency
N = carrier density
Q = carrier charge
m = carrier mass
ϵ_0 = permittivity of free space

To maximize positive ion oscillations, the tube should be tuned to resonate the heavy ion plasma frequency. Since the plasma in the tube acts as inductance,[11] the tube can be tuned by adding a large capacitance in parallel with it (figure 5).

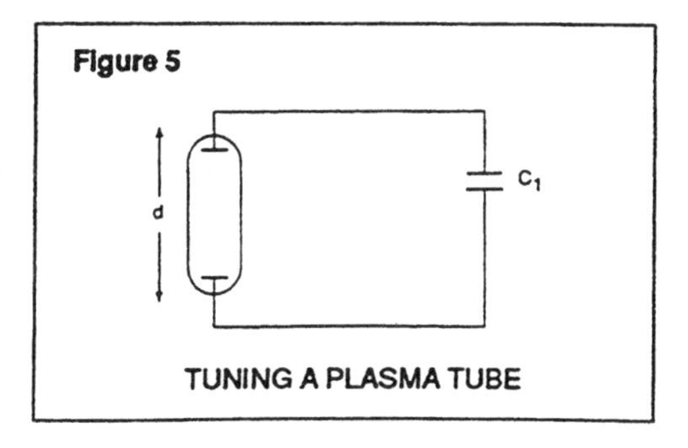

TUNING A PLASMA TUBE

In this circuit

$$L = \frac{dm}{SNQ^2}$$

where

L = equivalent inductance
d = electrode separation
S = electrode area
C_1 = added capacitance
$C_0 = \epsilon_0 \frac{S}{d}$ = electrode vacuum capacitance

Such a circuit will resonate at

$$\omega = \sqrt{\frac{1}{L(C_0+C_1)}} = \sqrt{\frac{SNQ^2}{dm(C_0+C_1)}} = \sqrt{\frac{NQ^2}{\epsilon_0 m(1+C_1/C_0)}} = \frac{\omega_{p-}}{\sqrt{1+C_1/C_0}}$$

where ω_{p-} is the electron plasma frequency.

For $C_1 >> C_0$,

$$\omega = \omega_{p-}\sqrt{C_0/C_1}$$

The heavy ion oscillations are maximized when the circuit resonates at the heavy ion plasma frequency ω_{p+}. This occurs when the tube is tuned by adding capacitance C_1 such that:

$$C_1 = C_0\left(\frac{\omega_{p-}}{\omega_{p+}}\right)^2$$

The plasma in the tube should be maintained by a minimum amount of energy, for if the tube gets too hot the turbulent collisions will disrupt the heavy ion oscillations. Moray appears to have used a small amount of radioactive material to maintain the plasma in some of his cold cathode ionic oscillation tubes.

Because the heavy ions maintain their oscillations in these tubes, they are able to absorb small pulses of energy when these pulses are applied in phase with the oscillations. This absorption will make the oscillations grow in amplitude. Thus small amounts of energy, when properly fed into the system, can integrate over time into large powerful oscillations.

Once a medium for lossless resonance is established, it is possible to absorb and then transduce high frequency electrical energy down the spectrum. The previously described "spectral diode" can be electrically realized if the ideal oscillators in the circuit are properly tuned plasma tubes. Many stages can be cascaded together. If the frequency ratio between the stages is three to one, this system would maximize the excitation of its lower frequency modes, and should optimally draw high-frequency energy down the spectrum.

TAPPING THE ZERO-POINT ENERGY?

The "spectral diode" draws high frequency energy from the environment irreversibly. (Most systems reach a point of equilibrium at any particular mode with the environment.) Does nature replenish those drained modes? If the underlying substrate of space is nonlinear in its electrical activity, the energy of even higher frequencies could spill down to replenish the depleted modes. But what would ultimately be the source of this very high frequency energy? It could be the all-pervading energy of space, the zero-point energy.[12-18]

Boyer[19] derives the spectrum of the zero-point energy by requiring it to be Lorentz invariant. The zero-point spectral density function is

$$\rho\,(\omega) = \frac{\hbar\omega^3}{2\pi^2 c^3}$$

Note the energy content approaches infinity as the frequency approaches infinity. If lower frequency modes started to become depleted, energy from the higher frequency modes would have to shift down the spectrum in order to maintain the required Lorentz invariance. Could this be the true action of the fabric of space? Does "nature abhor a vacuum" even in the frequency domain? Need all we do to tap the limitless high frequency modes of the zero-point energy is irreversibly draw energy down the spectrum?

If so, the "spectral diode" could provide limitless energy as long as the low frequency (output) mode does not saturate.

SUMMARY

The environment contains tremendous amounts of high frequency energy. But if this is so, why can't today's standard detectors and field strength meters detect it? It is because these detectors are in thermodynamic equilibrium with the environment. Most of the energy absorbed is reradiated back into the environment, often as (infrared) heat. Only a small net rectified energy drives the detector. There is no coherent transduction of the energy down the spectrum. (In fact, it is transduced the other way.) A system capable of irreversibly transducing the frequency of the energy must be a nonlinear system, well away from thermodynamic equilibrium. Such systems have been identified by Nicholis, Prigogine[20] and Haken[21] to exhibit self-organizing properties. Such a system can draw in the very high frequency energy from the environment and convert it into large amplitude, low frequency energy. If nature replenishes the high frequency modes, then the "spectral diode" could tap a limitless supply of energy.

ACKNOWLEDGEMENTS

The author wishes to thank H. Roy Curtin and David L. Faust for helpful discussions.

REFERENCES

1. F. Bueche, *Introduction to Physics for Scientists and Engineers*, McGraw Hill, 1969, p. 612.

2. X. C. Rebbi, "Solitons," *Sci. Amr.*, 92 (Feb. 1979).

3. J. M. Manley, H. E. Rowe, "Some General Properties of Nonlinear Elements Part I. General Energy Relations," Proc. IRE 44, 904 (1956).

4. M. T. Weiss, "Quantum Derivation of Energy Relations Analogous to Those for Nonlinear Reactances," Proc. IRE 45, 1012 (1957).

5. J. Brown, "Proof of the Manley-Rowe Relations from Quantum Considerations," *Electron. Lett.* 1, 23 (1965).

6. P. A. Sturrock, "Action Transfer and Frequency-Shift Relations in the Nonlinear Theory of Waves and Oscillations," *Ann. Phys.* 9, 422 (1960).

7. P. Penfield Jr., *Frequency-Power Formulas,* Wiley, N.Y. (1960).

8. A. C. Scott, F. Y. F. Chu, D. W. McLaughlin, "The Soliton: A New Concept in Applied Science," Proc. I.E.E.E. 61, No. 10, 1443 (Oct. 1973).

9. A. H. Nayfeh, D. T. Mook, *Nonlinear Oscillations,* Wiley, N.Y. (1979) p. 423.

10. T. H. Moray, *The Sea of Energy,* Cosray Research Institute, Salt Lake City (1978).

11. P. Lorrain, D. Corson, *Electromagnetic Fields and Waves,* Freeman Co. (1970) p. 485.

12. T. H. Boyer, "Random Electrodynamics: The Theory of Classical Electrodynamics with Classical Electromagnetic Zero-Point Radiation," *Phys. Rev.* D 11, No. 4, 790 (1975).

13. C. Cercignani, L. Galgani, A. Scotti, "Zero-Point Energy in Classical Non-Linear Mechanics," *Phys. Rev. Lett.* 38A, No. 6, 403 (1972).

14. T. W. Marshall, "Statistical Electrodynamics," Proc. Cambr. Phil. Soc. 61, 537 (1965).

15. E. G. Harris, *A Pedestrian Approach to Quantum Field Theory,* Wiley (1972); Chap. 10: "The Problem of Infinities in Quantum Electrodynamics."

16. H. B. G. Casmir, "Introductory Remarks on Quantum Electrodynamics," *Physica* 19, 846 (1953).

17. M. Ruderfer, "Neutrino Structure of the Ether," *Lett. Al Nuovo Cimento* 13, No. 1, 9 (1975).

18. C. Lanczos, "Matter Waves and Electricity," *Phys. Rev.* 61, 713 (1942).

19. T. H. Boyer, "Derivation of the Blackbody Radiation Spectrum Without Quantum Assumptions," *Phys. Rev.* 182, No. 5, 1375 (1969).

20. G. Nicolis, I. Prigogine, *Self-Organization in Nonequilibrium Systems,* Wiley, N.Y. (1977).

21. H. Haken, *Synergetics,* Springer-Verlag, N.Y. (1971).

NOISE AS A SOURCE OF ENERGY

December 1982

ABSTRACT

The possible use of semiconductor noise as an energy source is explored. The nonlinear behavior of the crystal's traps, microplasma and exciton surface plasma may allow self-organizing behavior and a wide-band absorption of energy. The detector of T. Henry Moray appears to have absorbed very high frequency sweeper emissions.

INTRODUCTION

Can noise be tapped as a feasible source of energy? At first glance this would seem to violate the second law of thermodynamics. We are asking random energetic events to cohere or organize themselves. A closed linear system at equilibrium could not do this. However, Nicolis, Prigogine[1] and Haken[2] identify systems which exhibit self-organizing behavior. These are open, nonlinear systems well away from equilibrium which are maintained by an energy flux through them. The question is can such a system interact with and organize random energy so that the noise itself becomes the required energy flux? If such a boot-strap behavior became feasible, then in principle it could be possible to tap the ultimate substrate of environmental noise: The zero point energy.[3-6]

It is well known that small amounts of noise energy from the environment can be rectified onto a capacitor. At some later time the energy can be used for work. Note that only a small amount of energy can be stored, for the environmental noise pulse must exceed the voltage already stored on the capacitor in order to be absorbed. Clearly, most of the noise energy is unused by such a simple system. However, many repetitions of this circuit (Figure 1) could be used to accumulate some energy. Yater[7-10] shows that by working with banks of rectifying circuits on a small scale, appreciable energy can be absorbed. Yater points out that by working on a microscopic scale, the rectification process is aided by quantum mechanical and near field effects, making his noise absorber all the more efficient. Yater's premise may be further supported by examining the electrical noise properties of semiconductor crystals where rectification-like energy storage occurs on an atomic scale.

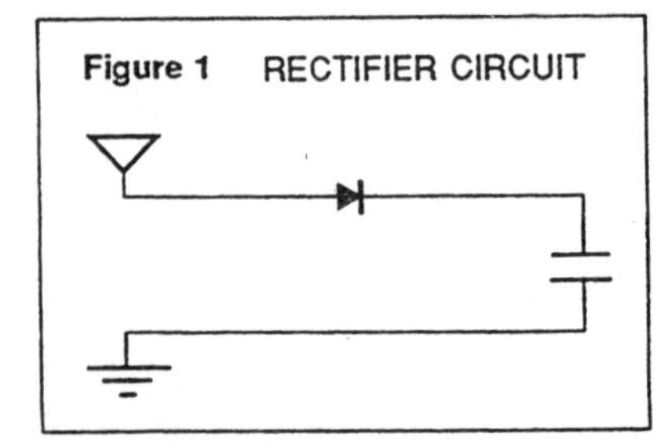
Figure 1 RECTIFIER CIRCUIT

SEMI-CONDUCTOR NOISE

In addition to the standard resistor thermal noise, semiconductors exhibit burst noise and 1/f noise. Many investigators[11-19] have concluded that burst and 1/f noise are associated with electron-hole traps near crystal defects or on the crystal's surface. These traps act as storage areas perhaps analogous to Yater's microscopic capacitors. Sometimes energy can be stored for long periods. Firle and Winston[20] observed continuation of the 1/f noise spectrum down to 6×10^{-5} Hz! However, much of the time the energy is released by electron hole recombination giving rise to burst-recombination noise. If a retrapping effect occurs as Wallick[21] suggests, then the general system conditions as recently identified by Keshner[22] would be fulfilled for a system to exhibit 1/f noise. These conditions are similar to the self-organizing conditions identified by Haken and Prigogine, i.e., a non-stationary, non-equilibrium, non-Markovian evolutionary system. For semi-conductors, a "flux" current is required to exhibit 1/f and burst noise. Miller[23] has suggested a phonon flux to explain 1/f noise in metal films. Note an energy flux of some sort is required to maintain the non-equilibrium, non-stationary system characteristic, and conversely Keshner has shown that if such systems contain sufficient correlative memory (at least one state variable per decade of frequency) they naturally exhibit 1/f noise.

For semi-conductors this "memory" is provided by the traps. Burgess[24] suggests that each trap communicates with a whole spectrum of energy levels in the conduction and valence bands. Shockley[25] points out that the ability of a recombination center to absorb or emit electron-hole pairs can be modulated by action at an adjacent trap. Hsu, et al.[26] show how a recombination center can modulate current through a nearby defect, and McWhorter[27] points out the non-independent behavior of closely spaced traps. Also Sikula, et al.[28] propose a three state, non-Markovian process relating generation-recombination through traps. There is ample evidence for correlative effects in the trapping mechanism of semi-conductors. Instead of being totally released as bursts, some energy can be retrapped and stored for even longer periods giving rise to the 1/f spectrum.

Strasilla, et al.[29] and Conti, et al.[30] stress the common conditions that give rise to both 1/f and burst noise, and their work supports the current modulation model of Hsu. To cohere burst noise using Hsu's model requires gaining control of the electrons trapped in the recombination center so that they all modulate the adjacent defect's barrier potential in phase. It is not clear how to gain control to this microscopic level. However, there is another type of burst noise that exists in back biased p n junctions on the threshold of avalanche breakdown. Rose[31] describes this condition as a microplasma. McKay[32] describes it as "onset noise" just before breakdown. Hsu observes that the microplasma bursts are a thousand times larger than the forward biased "modulation" bursts. Note this microplasma state is similar to biasing a gas discharge junction into the Townsend region.[33] It is in this region where valve rectifying action is maximized. In addition, at this point may occur a possible zero-point energy coherence by the change in state from bound charge to microplasma. Boyer[3] describes how matter can affect the zero-point energy and Rauscher[34] describes how a plasma and the zero-point energy interact. It is in this threshold region where a small trigger pulse can release the stored energy in the junction resulting in an avalanche charge multiplication process.

If noise energy from the ambient can accumulate in a back biased pn junction through a Yater-like microscopic rectification (perhaps into traps), a natural, cohering, release mechanism is available through the avalanche process.

THE MORAY VALVE

In the 1930's Moray[35] developed a solid state amplifying valve that resembled the point contact transistor. Moray powered his valve by doping his germanium with radioactive material. McKay[36] used alpha bombardment to induce charge multiplication in his study of the avalanche process. Moray's fissionable material not only helped maintain an internal microplasma, but it produced a surface plasma and air plasma as well.

The components of one of Moray's valves were as follows (Figure 2):

Germanium (Ge) was doped with n type impurities as well as fissionable material. In point contact transistors n type impurities tend to accumulate holes near the collector which lead to current multiplication.[37]

Molybdenum (Mo) and molybdenum sulfide (MoS) were used for the point contact pin. Molybdenum is a p type material thus enhancing the p n junction. The rectification occurred predominantly at this point-contact. Note that Plaksii, et al.[38] and Luque, et al.[39] have observed enhanced noise at point contact junctions. If this noise energy could be absorbed, the amplifying properties of the valve could be enhanced.

Figure 2

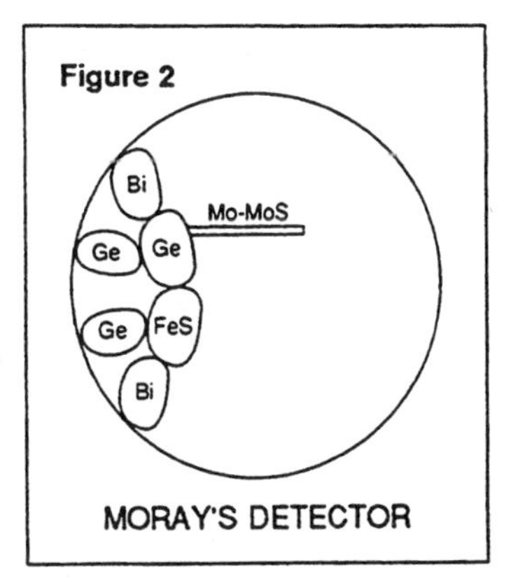

MORAY'S DETECTOR

Iron Sulfide (FeS) has traditionally been used in the original crystal set radio receivers. It is a p type semi-conductor.[40] Metallic sulfides are known for their surface luminescence and exciton formation.[41]

Excitons may provide the cohering mechanism for noise absorption. Redfield[42] notes that excitons are drawn toward crystal defects and energy can be exchanged by florescence. Simpson[43] observes that excitons can defuse from one crystal to a contiguous crystal. This would allow an energy exchange throughout the germanium and iron sulfide surface plasmas.

The excitons can be likened to a soliton form. Bostick, et al.[44] suggests that solitons can absorb and feed on "small waves" (e.g., noise) that they encounter. If these excitons are then drawn toward the crystal surface defect, a type of negentropic condition may be set up where the excitons sweep energy toward the point contact where their energy can be released synchronously with a triggering current pulse. Excitons can also provide the medium of energy transport between the air ions (in the cold plasma produced by the radioactivity) and the internal p n microplasma.

Moray used moist vapor to aid the ionic action of the air plasma[45] and to enhance the surface plasma—air plasma interface as well. It is interesting to note that Pearson, et al.[46] observed that 1/f noise was enhanced by orders of magnitude when water vapor was in the ambient of the p n junction. McWhorter[47] noticed that the 1/f noise is enhanced by traps from ions on the oxide surface of the crystal. It is conceivable for these effects to combine with the exciton noise gathering action to augment the amount of noise energy rectified by the valve.

Bismuth (Bi) is predominantly used as an electron (n type) donor to the surface plasmas on the germanium and iron sulfide pellets. Moray stressed utilization of cold cathodes in his work and bismuth is an excellent material to aid this action by electron donation for exciton formation. The excitons could then be triggered to release their electrons from the surface into the surrounding medium utilizing the alpha bombardment as the energy source. The release would be controlled by the signal on the point contact pin. It is interesting to note that Moray was working with the appropriate solid state materials well before any discussions of the transistor appeared in the literature.

A hemispherical metallic cavity was used to encase the pellets. (Figure 3.)

Figure 3

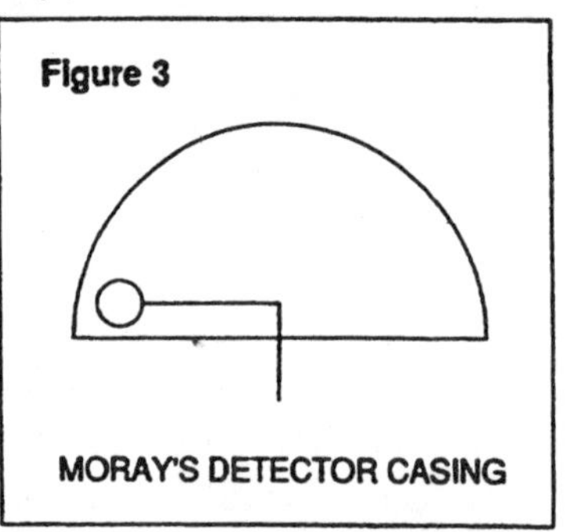

MORAY'S DETECTOR CASING

Both the iron sulfide and germanium pellets are insulated from the metal case making electrical contact through the bismuth. The pin constituted one electrical contact while the case itself was the other. The apparatus acted as an amplifying diode. Moray stressed the importance of the hemispherical casing.[54] It provides a resonant cavity for a wave that rapidly changes frequency, i.e., a "chirping" waveform. Note near the pellet, where the air ionization is the greatest (due to the radioactivity), the ion density is the largest, begetting a large characteristic plasma frequency. The pellet is mounted near the edge where the vertical distance is small. Toward the center of the cavity, the standing wave distance is larger and the ionization is less giving a lower characteristic frequency. Thus, this detector shell is designed to resonate with a sweeping or chirping waveform.

Moray claimed the radiant energy intercepted by his detector would surge in at regular intervals. Gerson, et al.[47] describe detecting sweeping emissions that seem to be very similar to what Moray was detecting:

> "Wideband noise bursts termed sweepers drift in frequency through portions of the HF and VHF bands. They are of two broad types: (1) instantaneous, and (2) drifting mainly from higher to lower frequencies. They are readily observed at many locations over the planet. Their occurrence maximizes between 24-26MHz. The instantaneous type probably is associated with thunderstorm activity. The drifting type may occur in trains that persist for hours. Individual members recur at closely the same time interval and display no significant dispersion. Intensities may be very high. They are generally not noticed when narrow band receivers are used. Their origin is not clear. . . ."

Their origin could be from ionospheric, magnetospheric, solar or other plasmas. Note this description is similar to the observation of whistlers or whistler wave solitons[48] except that whistlers are observed at a lower frequency (on the order of 100KHz). Could Moray have detected sweeper type emissions except at a higher (and considerably more energetic) frequency? Note the importance of a broad band detector. Narrow band receivers would only see it as noise and could never synchronously absorb a rapidly chirping high frequency waveform.

CONCLUSION

It appears that the shot, flicker (1/f), recombination, burst, and microplasma noise in semi-conductors have their power originate from the external current flowing through the crystal. The traps and crystal defects provide the storage and modulation mechanisms needed to give rise to these types of noise. These noise types in and of themselves do not appear to be independent sources of energy. They actually constitute an incoherency loss to the original external current applied to the crystal.

On the other hand, the traps constitute an energy storage capability that perhaps can be utilized. Excitons negentropically gravitate toward the traps and the floresence experiments indicate that energy can be stored for long periods. This energy may interact with the three plasmas associated with a semi-conductor crystal: The internal microplasma, the exciton surface plasma and the external ambient ions (e.g., gas plasma). The plasmas do provide a quivering, nonlinear medium capable of supporting solitons and vortices. (Note the vortex is considered a soliton form[49,50]). It is in the nonlinear behavior of the plasma where the self-organization and the possibility of noise absorption and coherence lies. The standard literature has not yet experimentally demonstrated the efficient capture of external noise energy in semi-conductors, but most of the studies have dealt with examining the crystal's "modulation" of direct currents. Little work has been done on trying to find resonant modes in the trapping mechanisms under higher frequency excitation. Perhaps this would be a fruitful area of investigation as it is well known that plasmas have various resonant modes.

The detector of Moray appears to be tapping an impinging energy that is not thermal or white noise. The most corroborating descriptions in the standard literature appear to be the sweepers and whistlers. Both of these forms of energy probably arise from the action of plasmas (ionospheric or beyond), and this type of energy is not readily detected by narrow band receivers. If this waveform occurs at very high frequencies, there could be considerable energy imbedded in it. A synchronous frequency step-down technique would be required to efficiently extract it.[51] This seems to be exactly what Moray's radiant energy device did.

It is possible that the plasma itself can interact with another form of energy. Ruderfer[52] suggests neutrinos. An individual nucleus rarely interacts with a neutrino, but perhaps the plasma as a whole can macroscopically interact with neutrinos if the characteristic frequencies match. Future work will explore the hypothesis that there is an extra energetic aspect associated with the oscillations of a plasma's heavy ions. This extra aspect could be vigorous vacuum polarization waves in the zero-point Fermi sea or perhaps the matter waves of Lanczos.[53] The key to understanding Moray's radiant energy device lies in the oscillating action of the heavy ions in his plasma tubes. Nonlinear oscillating plasmas offer the best hope for the negentropic absorption of ambient energy.

ACKNOWLEDGEMENTS

The author wishes to thank David Faust and Josh Reynolds for many helpful discussions.

REFERENCES

1. G. Nicolis, I. Prigogine, *Self-Organization in Nonequilibrium Systems,* New York: Wiley, (1977).
2. H. Haken, *Synergetics,* New York: Springer-Verlag, (1971).
3. T. H. Boyer, "Random Electrodynamics: The Theory of Classical Electrodynamics with Classical Electromagnetic Zero-Point Radiation," *Phys. Rev.* D 11, No. 4, 790 (1975).
4. C. Cercignani, L. Galgani, A. Scotti, "Zero-Point Energy in Classical Non-Linear Mechanics," *Phys. Rev. Lett.* 38A, No. 6, 403 (1972).
5. T. W. Marshall, "Statistical Electrodynamics," *Proc. Cambr. Phil. Soc.* 61, 537 (1965).
6. H. B. G. Casmir, "Introductory Remarks on Quantum Electrodynamics," *Physica* 19, 846 (1953).
7. J. C. Yater, "Power Conversion of Energy Fluctuations," *Phys. Rev.* A 10, No. 4, 1361 (1974).
8. J. C. Yater, "Relation of the Second Law of Thermodynamics to the Power Conversion of Energy Fluctuations," *Phys. Rev.* A 20, No. 4, 1614 (1979).
9. J. C. Yater, "Physical Basis of Power Conversion of Energy Fluctuations," *Phys. Rev.* A 26, No. 1, (1982).
10. J. C. Yater, "Particle Interactions in the Power Conversion of Energy Fluctuations," (to be published).
11. G. Blasquez, "Excess Noise Sources Due to Defects in Forward Biased Junctions," *Solid State Electron.* 21, No. 11-12, 1425 (1978).

12. G. Blasquez, J. Caminade, "Physical Sources of Burst Noise," *Noise in Physical Systems,* New York: Springer-Verlag, 60 (1978), pg. 60.
13. G. Doblinger, "A Burst Noise Model for Integrated Bipolar Transistors with Anomalous IV Characteristics," *Noise in Physical Systems,* New York: Springer-Verlag, (1978), pg. 64.
14. A. G. Grant, A. M. White, B. Day, "Low Frequency Noise and Deep Traps in Shottky Barrier Diodes," *Noise in Physical Systems,* New York: Springer-Verlag, (1978), pg. 175.
15. K. B. Cook, A. J. Brodersen, "Physical Origins of Burst Noise in Transistors," *Solid State Electron.* 14, No. 12, 1237 (1971).
16. S. T. Hsu, R. J. Whittier, "Characteristics of Burst (Popcorn) Noise in Transistors and Operational Amplifiers," *International Electron Devices Meeting IEEE,* (1969), pg. 86.
17. J. C. Martin, D. Esteve, G. Blasquez, "Burst Noise in Silicon Planar Transistors," *Conference on Physical Aspects of Noise in Electronic Device,* London: P. Peregrinus LTD, (1968), pg. 99.
18. J. C. Martin, G. Blasquez, A. DeCacqueray, M. DeBrebisson, C. Schiller, "The Effect of Crystal Discolorations on Burst Noise in Silicon Bipolar Transistors," *Solid State Electron.* 15, No. 7, 739 (1972).
19. K. F. Knott, "Evidence of Collector-Base Junction Burst Noise," *Electron. Let.* 15, No. 6, 198 (1979).
20. T. E. Firle, H. Winston, *J. Appl. Phys.* 26, 716 (1955).
21. G. C. Wallick, "Size Effects in the Luminescence of ZnS Phosphors," *Phys. Rev.* 84, 375 (1951).
22. M. S. Keshner, "1/f Noise," *Proc. IEEE,* 70, No. 3, 212 (1982).
23. S. C. Miller, "1/f Noise from Surface Generation and Annihilation: Application to Metal Films," *Phys. Rev.* B 24, No. 6, 3008 (1981).
24. R. E. Burgess, *Brit. J. Appl. Phys.* 6, 185 (1955).
25. W. Shockley, *Electrons and Holes in Semiconductors,* D. Van Nostrand Co., (1950), pg. 342.
26. S. T. Hsu, R. J. Whittier, C. A. Mead, "Physical Model for Burst Noise in Semiconductor Devices," *Solid State Electron.* 13, 1055 (1970).

27. A. L. McWhorter, "1/f Noise and Germanium Surface Properties," *Semiconductor Surface Physics*, Philadelphia: University of Pennsylvania Press, (1957), pg. 207.

28. J. Sikula, M. Sikulova, P. Vasina, B. Koktavy, "Burst Noise in Diodes," *Sixth International Conference on Noise in Physical Systems*, (1981), pg. 100.

29. U. J Strasilla, M. J. O. Strutt, "Measurement of White and 1/f Noise within Burst Noise," *Proc. IEEE* 62, No 12, 1711 (1974).

30. M. Conti, G. Corda, "Identification and Characterization of Excess Noise Sources in ICS by Correlation Analysis," International Electron Devices Meeting Technical Digest IEEE, (1973), pg. 248.

31. D. J. Rose, "Microplasmas in Silicon," *Phys. Rev.* 105, No. 2, 413 (1957).

32. K. G. McKay, "Avalanche Breakdown in Silicon," *Phys. Rev.* 94, No. 4, 877 (1954).

33. L. B. Loeb, *Fundamental Processes of Electrical Discharge in Gases*, New York: John Wiley, Inc., (1939), pg. 372.

34. E. Rauscher, "Electron Interactions and Quantum Plasma Physics," *J. Plasma Phys. 2*, Part 4, 517 (1968).

35. T. H. Moray, *The Sea of Energy in Which the Earth Floats*, Salt Lake City: Cosray, 4th Edition, (1960), pg. 132.

36. K. G. McKay, K. B. McAfee, *Phys. Rev.* 91, 1079 (1953).

37. J. B. Arthur, A. F. Gibson, J. B. Gunn, "Carrier Accumulation in Germanium," *Proc. Phys. Soc.* Section B, Vol. 169, Part 7, No. 439B, 697 (1956).

38. V. T. Plaksii, A. P. Zakharov, V. N. Svetlichnyi, V. V. Starostenko, "High Frequency Noise of Point Contact Between Metal and Semiconductor," *IZV. VUZ Radioelektron* (USSR) Vol. 15, No. 5, 657 (1972). Translation in *Radio Electron. and Comm. Syst.* (USA).

39. A. Luque, J. Mulet, J. Rodriguez, R. Segovia, "Proposed Dislocation Theory of Burst Noise in Planar Transistors," *Electron. Lett.* 6, No. 6, 176 (1970).

40. J. R. Gosselin, M. G. Townsend, R. J. Tremblay, "Electric Anomalies at the Phase Transition in FeS," *Solid State Comm.* 19, 799 (1976).

41. J. J. Lambe, C. C. Klick, D. L. Dexter, "Nature of Edge Emission in Cadmium Sulfide."

42. D. Redfield, "A Mechanism for Energy Transport by Excitons," Proc. International Conference on Phys. Semiconductors, Kyoto, (1966), page 139, *J. Phys. Soc.* Japan 21, Supp. (1966).

43. O. Simpson, "Electronic Properties of Aromatic Hydrocarbons III. Diffusion of Excitons," *Proc. Royal Soc.* A238, 402 (1957).

44. W. H. Bostick, O. S. F. Zuker, "Theoretical and Practical Aspects of Energy Storage and Compression," *Energy Storage, Compression, and Switching,* New York: Plenum, (1976), page 71.

45. T. H. Moray, "Electrotheurapeutic Apparatus," U.S. Patent No. 2,460,707, (1949).

46. G. L. Pearson, H. C. Montgomery, W. L. Feldman, *J. Appl. Phys.* 27, 91 (1956).

47. W. C. Gerson, W. H. Gossard, "Sweeping Emissions," *Phys. Canada* 27, No. 4, 39 (1971).

48. M. Watanabe, "On the Whistler Wave Solitons," *J. Phys. Soc.* Japan 45, No. 1, 260 (1978).

49. G. L. Lamb, "Solitons and the Motion of Helical Curves," *Phys. Rev. Lett.* 37, No. 5, 235 (1976).

50. F. Lund, T. Regge, "Unified Approach to Strings and Vortices with Soliton Solutions," *Phys. Rev.* D14, No. 6, 1524 (1976).

51. M. B. King, "Stepping Down High Frequency Energy," *Proceedings of the First International Symposium on Non-Conventional Energy Technology,* Toronto, (1981), page 145.

52. M. Ruderfer, "Neutrino Structure of the Ether," *Lett. Al Nuovo Cimento* 13, No. 1, 9 (1975).

53. C. Lanczos, "Matter Waves and Electricity," *Phys. Rev.* 61, 713 (1942).

54. J. E. Moray, "Theory of Operation of Apparatus," *Sun Day Conference,* Philadelphia, (May 1978).

MACROSCOPIC VACUUM POLARIZATION

September 1984

ABSTRACT

Nikola Tesla and T. Henry Moray claimed inventions that apparently absorbed anomalously large amounts of environmental radiant energy. These inventions may have utilized a macroscopic vacuum polarization zero-point energy coherence associated with ion-acoustic oscillations in a plasma.

INTRODUCTION

It is generally taught to engineers that Maxwell's equations constitute a complete theoretical basis for all macroscopic electrodynamics at the engineering level. On the other hand it is also taught that other effects (e.g., pair production, vacuum polarization, zero-point fluctuations) can occur that are not predicted by Maxwell's equations, but these are at the "quantum level." It is tacitly implied that quantum vacuum effects do not relate to "engineered" systems. Yet in reality there is only one electrodynamics and to totally engineer it requires an understanding at its basis.

Quantum electrodynamics shows that the basis of all electrical phenomena is the vacuum, where tremendous fluctuations of electrical field energy occur; this energy is called the zero-point energy,[1-4] and by some consideration is the modern term for the ether. Tesla[63] and later Moray[60] believed their devices interacted directly with the ether. Today's physics recognizes that matter interacts with the vacuum zero-point energy.[1] The term commonly used to describe the interaction of charged particles with the vacuum is "vacuum polarization." Whereas, many investigators use the term only to mean pair production, here it shall include all the states of the vacuum from effects in the low field linear regime (where Maxwell's equations apply), through the nonlinear regime on the threshold of pair production, to the "bifurcation catastrophe"[5] regime where pair production occurs.

In this paper, literature is identified that shows that the zero-point energy is necessary in electrodynamics and the manner in which it is needed to explain the radiation of a uniformly accelerated charge. References are also cited that demonstrate that the various elementary particles interact differently with the zero-point vacuum fluctuations. It is suggested that as a result of this interaction, ions may have different radiation characteristics than conduction electrons. This motivates the hypothesis that the ion-acoustic mode of a plasma may produce a propagating, coherent, macroscopic vacuum polarization. A qualitative vacuum polarization model is proposed to explain why conduction electrons would not readily detect this type of radiant energy. It is suggested that the ion-acoustic oscillations in Tesla's and Moray's devices absorbed longitudinal, vacuum polarization displacement currents, and that these inventors actually did discover a novel form of environmental radiant energy.

UNIFORM ACCELERATION OF CHARGE

Any complete theory of electromagnetism must include the zero-point energy, for ignoring it leads to contradictions and paradoxes at a very fundamental level. One difficulty in electrodynamics is known as the equivalence paradox. Here a uniformly accelerated charge is recognized to radiate. However, a charge suspended at rest in a uniform gravitational field does not. According to general relativity, a uniformly accelerating system in free space should be equivalent to one at rest in a uniform gravitational field. Thus, in this case the principle of equivalence seems to be violated. This problem has been discussed in the literature at the classical level without adequate resolution. For example, Rolrhich,[6] Atwater,[7] and Ginzburg[8] conclude that radiation is a function of the acceleration of the observer in relation to the source charge. But as Ginzburg asks, what are photons, and what propagates at the velocity of light if it can be made to appear or disappear depending on the acceleration of the observer? Boulware[9] similarly suggests that "the way out of the paradox is to deny that the concept of radiation is the same in the accelerated and unaccelerated frames."[9] This interpretation likewise throws out the independent existence of light by linking it to the motion of the observer.

C. M. Dewitt[10] and B. S. Dewitt[11] acknowledge the violation of equivalence but state that electric charge is an "unfair," i.e., scientifically invalid, test of the principle of equivalence since a real gravitational field is only uniform locally, but a charge's field persists to infinity. B. S. Dewitt[11] also states that spinning neutral bodies also deviate from geodesic motion and they too are "unfair" tests of the principle of equivalence since angular momentum is a manifestation of a "nonlocal" phenomenon. It appears that either one must admit some of the laws of physics disobey the principle of the equivalence or that light cannot exist as an independent entity. The equivalence paradox has not yet been adequately resolved at the classical level.

An even more basic problem appears in classical electrodynamics regarding the uniform acceleration of charge. It is generally accepted that the radiated power is proportional to the square of the acceleration as computed by the Larmor formula:[12]

$$p = \frac{2}{3}\frac{e^2 a^2}{c^3} \qquad (1)$$

Yet the radiation reaction friction force experienced by the charge is proportional to the first time derivative of the acceleration:[12]

$$F = \frac{2}{3}\frac{e^2}{c^3}\frac{da}{dt} \qquad (2)$$

For uniform acceleration this derivative is zero while the acceleration is not. The particle radiates but does not lose kinetic energy. Where does the radiation energy come from? Fulton,[13] Ginzburg,[8] and Boulware[9] conclude it has to come from the charge particle's source field. But Pauli,[14] Vasudevam,[15] and Luiz[16] come to the opposite conclusion, stating, "...we cannot accept the assumption of radiation from the charge because otherwise the internal energy of the particle should be exhausted."[16] Luiz further argues, "The law of action and reaction is a fundamental law in physics and if the radiation reaction is zero, certainly there is no radiation."[16] Ginzburg[8] points out that this fundamental problem manifests itself in attempts to match theory with experiments measuring synchrotron radiation. Surprisingly, Leibovitz[12] concludes that Maxwell's equations are incompatible with uniform accelerating motion! Electrodynamics clearly suffers serious problems at the classical level where the zero-point vacuum fluctuations are ignored.

ZERO-POINT ENERGY

When the formalism for the zero-point energy is introduced, some of these issues may be better understood and resolved. Callen[18] demonstrates that the vacuum fluctuations manifest themselves even at a classical level: "The existence of a radiation impedance for the electromagnetic radiation from an oscillating charge is shown to imply a fluctuating electric field in the vacuum."[18] Candelas[19] shows that "...pressure fluctuations associated with these energy fluctuations would confer on the charge an irregular motion. This motion would represent a non-constant acceleration and so would also lead to a systematic reaction dampening force acting on the charge."[19] This supports Fulton's ad hoc suggestion that hyperbolic motion is unphysical.[15] Sciama concludes, "...the classical results regarding the radiation emitted by an electron and the radiative reaction force on an electron...can be understood in terms of the spectrum of the field fluctuations perceived by the charge."[12] Sciama suggests a borrowing mechanism where the radiated energy is borrowed from the vacuum field during periods of uniform acceleration and then given back during nonuniform acceleration.[12] Thus any consideration of the nature of a charged particle and of radiation production should include the charge's interaction with the zero-point energy. The resolution of the equivalence paradox may come with the development of quantum gravity theories[20] in which the zero-point energy plays a crucial role.

Any complete theory of electrodynamics must include the zero-point vacuum fluctuations and their interaction with matter. Boyer[1] shows that matter affects the zero-point fluctuations and they in turn feed back and affect matter. In fact, it appears that elementary particles can be viewed as organized coherences or spatial resonances in the zero-point sea.[21] Senitzky[22] shows that the charge's source field and the vacuum fluctuations are inseparably intertwined as "merely two sides of the same quantum mechanical coin."[22] This duality is likened to quantum mechanics' wave-particle duality. Sciama further supports this idea of charge vacuum synergy by noting that "it is not in general possible to divide the stress energy tensor into a 'real particle part' and a 'vacuum polarization part' in an unambiguous way."[12]

VACUUM POLARIZATION

Different types of particles interact differently with the vacuum zero-point energy.[23-25] In a first order model, protons, nuclei, and heavy ions generally produce a spherically distributed vacuum polarization with lines of polarization converging sharply to the particle (Figure 1). In fact, Greenberg[26] has demonstrated that if the nucleus becomes large enough, the intensity of the polarization precipitates real electron-positron pairs from the vacuum.

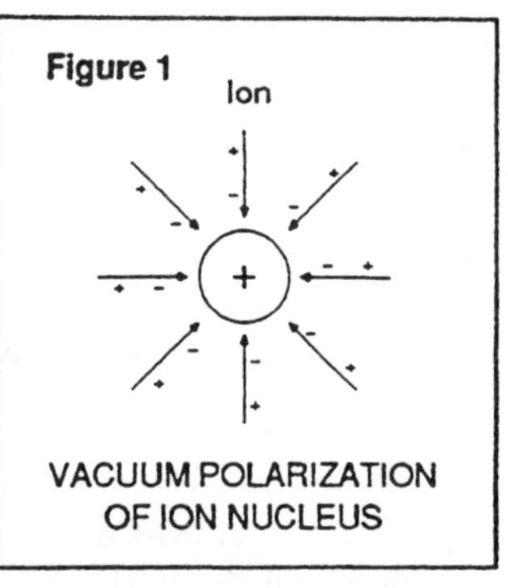

Figure 1
VACUUM POLARIZATION OF ION NUCLEUS

The spin of a particle also affects the vacuum fluctuations. Sciama notes that charged fields of differing spins (0, ½, 1) give rise to different vacuum states.[12] Vorticity can also appear in the vacuum. Graham[27] has experimentally observed a macroscopic vacuum angular momentum caused by a static electromagnetic field's circulating Poynting vector. It is clear that different particles give rise to different vacuum interactions.

In view of this, can a conduction electron radiate differently than an ion? The quantum mechanical wave function description of the electron in matter or in a conductor is that of a smeared charge cloud. This smearing dilutes the vacuum polarization intensity and prevents the lines of polarization from converging onto the electron in a stable, orderly way (Figure 2). The electron could be described as a light, "ethereal" particle whose interactions with the zero-point fluctuations alter its form and actually cause it to smear. This intertwining interaction helps explain atomic ground state stability.[19] The electron tends to stabilize into "standing wave" harmonic eigenstates in matter. This smeared cloud is in

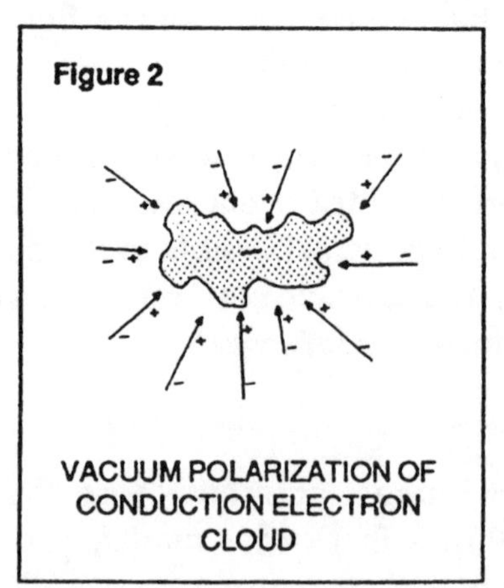
Figure 2
VACUUM POLARIZATION OF CONDUCTION ELECTRON CLOUD

equilibrium with the zero-point fluctuations. If we postulate there exist in the environment vacuum polarization displacement currents that can follow a particle's lines of polarization, then these currents would converge onto an ion but not converge onto a smeared electron cloud. Any environmental vacuum polarization displacement current would pass right through the smeared, fluctuating electron cloud. Senitzky shows that the "vacuum field plays no [net] role when the atomic system is an harmonic oscillator"[22] and that "linear oscillators such as antennas cannot in principle experience the effect of the vacuum field."[22] Also, Sciama shows that "for an [electron-based] detector at rest, the excitations caused by these zero-point fluctuations are precisely cancelled by its spontaneous emission rate."[12] Thus, the smeared electron cloud maintains thermodynamic equilibrium with the vacuum and could not absorb zero-point vacuum polarization surges.

However, the concentrated mass of the nucleus or heavy ion could interact with vacuum polarization modes, for its own vacuum polarized field convergently channels the longitudinal oscillations directly to the particle, *altering its momentum*. Note the two-way effect: the heavy particle can induce a spherically symmetric and convergent vacuum polarization around itself. As a transmitter, this field then launches vacuum polarization displacement currents whenever the particle moves or oscillates; as a receiver this field tends to channel those oscillations convergently onto the spherical particle, altering its motion. Because the heavy ion can maintain spherically convergent, stable lines of polarization, it becomes a transducer for transmitting and detecting longitudinal, vacuum polarization propagation modes that a conduction electron could not respond to and therefore could not detect.

Vacuum polarization effects can become powerfully synergistic when more than one ion or nucleus is involved. Roesel[28] describes the vacuum polarization potential for two extended charge distributions, and Soff[29] describes the "shake-off of the vacuum polarization cloud" in heavy ion collisions as a "collective type of electron-positron creation due to coherent action...."[29] Rauscher[30] was perhaps the first to suggest that coherent, vacuum quantum electrodynamic effects could take place in a plasma by demonstrating that vacuum polarization makes a significant contribution to the plasma's effective permittivity and conductivity. The nonlinear vacuum polarization description for a conglomerate of oscillating heavy ions would be quite complex and not readily solvable by the standard renormalization techniques. Modeling on a magnetohydrodynamic level would be more appropriate. In plasma analysis, the oscillations in polarization modulate the effective permittivity.[30] If similar modeling is applied to the vacuum's zero-point activity, the macroscopic, longitudinal, vacuum polarization oscillations could be described as "permittivity waves." A plasma model for the zero-point activity may yield a reasonable approximation, since Melrose[32] shows that the "vacuum polarization tensor in the presence of strong static homogeneous magnetic fields...reduces to forms equivalent to the magneto-acoustic and shear Alfven modes in a plasma."[32] Such modeling could predict longitudinal propagation modes. This could be reasonable, since Cover[31] demonstrates that vacuum polarization can give rise to longitudinal photon-like resonances. In this model, the highly nonlinear description of a group of ions interacting

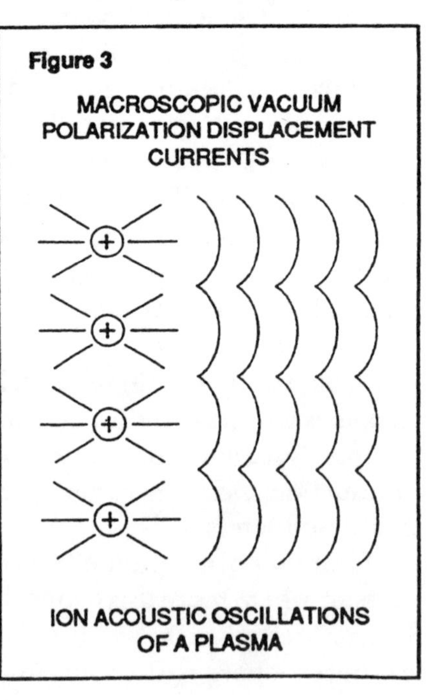

with the vacuum energy could fulfill the nonequilibrium conditions identified by Nicholis, Prigogine,[33] and Haken[34] that give rise to self-organizing coherent behavior. If the ions in a plasma synchronously oscillate together, the intense vacuum polarization associated with the individual ions could coherently and synergistically add to give rise to a very intense macroscopic vacuum polarization (Figure 3).

ION-ACOUSTIC OSCILLATIONS

A natural place to look for evidence of macroscopic vacuum polarization would be in a plasma. The coherent oscillation of plasma ions is known as the ion-acoustic mode. Many investigators have observed that the ion-acoustic mode is associated with large radiant energy absorption,[35-37] vigorous high-frequency spikes,[38-42] runaway electrons,[43,44] rapid and anomalous plasma heating,[45-48] and anomalous plasma resistance.[48-51] Could any of the ion-acoustic anomalies be associated with the existence of macroscopic vacuum polarization effects?

There is evidence for ion-acoustic activity in nature's plasma. The evidence comes indirectly from the observation of whistler or sweeper waveforms. Whistlers[55] are waveforms that rapidly downshift their frequency and cannot normally be detected by standard narrow-band receivers. They are observed with increased ion-acoustic activity in laboratory plasmas.[38,51-54] Sweeping emissions are also observed in nature.[56-58] The following description of sweepers by Gerson[56] is similar to what Moray[60] described as the source of radiant energy driving his invention:

> "Wideband noise bursts termed sweepers drift in frequency through portions of the HF and VHF bands. There are two broad types: (1) instantaneous, and (2) drifting mainly from higher to lower frequencies. They are readily observed at many locations over the planet. Their occurrence maximizes between 24–26 MHz. The instantaneous type is probably associated with thunderstorm activity. The drifting type may occur in trains that persist for hours. Individual mem-

> bers recur at closely the same time interval and display no significant dispersion. Intensities may be very high. They are generally not noticed when narrow band receivers are used. Their origin is not clear...."[56]

These sweepers may originate from tropospheric, ionospheric, magnetospheric, exospheric, and solar plasmas.[57,58] Podgornyi[59] notes that "the interplanetary medium is a giant reservoir filled with plasma in which various phenomena connected with collective interactions take place."[59] Webb[58] shows this activity persists through the atmosphere: "The geoenvironment is permeated with an electrical structure and with active electrical processes which serve to unify and control geoelectricity and to inter-relate geoelectricity with other physical aspects of the earth and its solar environment."[58] If the sweepers are associated with the vigorous ion-acoustic activity, then the action of the atmospheric, magnetospheric, exospheric, interplanetary and solar plasmas could be a source of longitudinal vacuum polarization displacement current. Could the ion-acoustic oscillations emit and detect this hypothesized form of radiant energy while conduction electrons not readily detect it?

MORAY'S DETECTOR

The hypothesis of the preceding section is viable, for the experiments conducted to confirm our present knowledge of electromagnetism has always used electron-based detectors. There is an exception, however. T. Henry Moray[60] experimented with ion oscillators and detectors, and as a result, he may have discovered what appears to be a novel energy source. Moray built a system of plasma tubes[61] and valves that were apparently tuned such that each tube resonated at its own ion-acoustic plasma frequency. The tubes were tuned and the circuit switching timed to shift the energy from the high-frequency stages down to the lower-frequency tubes.[62] A feature of resonating the ion-acoustic mode would be that the individual ions can experience a mutually coherent, low-loss harmonic motion without being totally disrupted by collisions. This would allow small pulses of energy from the previous stage to synchronously augment the oscillations. (Electrons are poor carriers for this purpose, since they are so light and their displacement is so large that they would undergo an excessive number of collisions per cycle of oscillation.) The ion-acoustic mode can also yield high electrical capacitance in each tube at its operating frequency. The oscillating ions in the plasma beget a maximal effective dielectric polarizability, while the anomalous resistance associated with the ion-acoustic mode prevents the plasma "dielectric" from breaking down. The use of many coupled stages at different operating frequencies allows a broadband interaction with the environmental energy. If impinging surges of vacuum polarization displacement currents encounter the ions in the tubes, the ions could synchronously move with them. Thus, Moray's ion-acoustic oscillators could resonate with the incoming vacuum polarization surges and absorb the energy.

Evidence to support the hypothesized existence of macroscopic vacuum polarization may come from studying the unusual characteristics of the output current from Moray's device. Most witnesses who observed the device in operation were impressed by the unusual, bright glow of the load-bank light bulbs. Another reported observation was that the conductive leads and the thin wires in the device remained cool even after hours of operation. This may be significant, for 30-gauge wire was used in the circuits within the device, and it delivered power on the order of kilowatts.[60] These observations might be explained by hypothesizing that the conductive leads acted as waveguides for the surrounding vacuum polarization displacement currents. In this case, the nuclei of the wire's metallic lattice would be the wave guiding structure with the conduction band electron cloud providing a smooth "continuity condition" to minimize scattering. Little net energy or momentum would be transferred to the conduction electrons since the vacuum polarization energy and the electron cloud are in thermodynamic equilibrium. The waveguide hypothesis can also be used to explain why a vigorous brush discharge was observed when the antenna was disconnected from the operating input-stage detector. Here, the detector itself set up high-frequency vacuum polarization displacement currents that were guided onto the antenna. This process would establish lines of polarization along the antenna that could then help channel environmental polarization displacement oscillations back to the detector's individual ions. This would augment the detector's effective cross-section for absorbing the environment's vacuum polarization energy. In this model, the ion oscillations and the vacuum polarization displacement currents are intimately phase-locked to yield a macroscopic wave-particle system. Perhaps Moray's invention was a manifestation of a macroscopic zero-point energy coherence.

The observations of the current from Moray's device suggest a qualitative experiment that could be used to help support the hypothesized significance of the ion-acoustic mode. A plasma tube is excited at its ion-acoustic frequency using an external power supply. If the hypothesized vacuum polarization displacement currents can be successfully coupled from the tube through conductive probes to an output load-bank circuit, then the current's characteristics can be compared to those that were observed in connection with Moray's device. If the output current during ion-acoustic resonance behaves similarly to the observed current from Moray's invention, and if the behavior cannot be duplicated by control tests using normal electrical conduction at the same power and frequency, then the tests would lend support to the hypothesis that the ion-acoustic mode launches macroscopic vacuum polarization displacement currents.

TESLA'S INVESTIGATIONS

The macroscopic vacuum polarization hypothesis may be applicable to explain Tesla's attempts to transmit and receive energy through high-potential devices (e.g., the towers at Colorado Springs[63] and Wardenclyffe[64]). The key transducer element in these structures would be the brush discharge corona around the sphere atop the towers. In order to allow coherent ion oscillations in this corona, it is most important to avoid sparking, for this will produce ion turbulence and disrupt the oscillations. Tesla avoided the sparking discharges by mounting smooth hemispherical capacitive structures at the top of his towers.[64] The circuit that energizes and couples to the corona would have to be tuned at the corona's ion-acoustic frequency. The tuning value is difficult to calculate, for the corona itself will increase the capacitance of the excitation circuit and alter its resonant frequency. Corona discharge studies[65-67] show that a stable, brush discharge corona can be induced by a unipolar radio frequency pulse burst. If the radio frequency matches the corona's ion-acoustic frequency, then a stable, coherent ion-acoustic oscillation can be maintained. This supports Corum's[68] suggestion that Tesla may have employed an X-ray ionization switch in order to achieve rectification of the driving radio frequency to induce and stabilize the corona atop his tower. If a stable ion-acoustic oscillation can be induced in the corona, then impinging vacuum polarization displacement currents could sympathetically couple to it and the energy could be absorbed into the coupled driving circuit.

Ion-acoustic oscillations may also be important for the production of ball lightning. Ball lightning[69,70] may be produced by replicating the bucking phase condition that Tesla associated with its production.[71] Ion-acoustic oscillations must first be induced in the corona around the Tesla coil. Then a signal or pulse must be abruptly switched into the circuit such that it is 180 degrees out of phase with the ion-acoustic oscillations. This bucking condition may induce a "vacuum polarization implosion" that could trigger the plasma to enfold into a vortex ring.[72-74] An anomalously long persistence of ball lightning triggered at relatively low energy levels could demonstrate the existence of macroscopic, coherent, self-organizing resonant states maintained by the zero-point energy.

SUMMARY

Many modern physicists have acknowledged that the zero-point energy or ether must be incorporated into any complete description of electromagnetic phenomena. Today's physics shows that different particles interact and polarize the vacuum in different ways. This suggests that ions can have different radiation characteristics than do conduction electrons. Even though today's physics cannot yet give solutions to the nonlinear multibody problem, it nonetheless recognizes the possibility that such nonlinear systems can manifest self-organizing coherent states. Since the individual ions in a plasma can coherently oscillate together, and since each ion exhibits an intense vacuum polarization, the ion-acoustic mode of a plasma may induce and detect macroscopic vacuum polarization displacement currents. The work of Moray and Tesla seems to support this hypothesis. It is hoped that this discussion will encourage experimental investigation of ion-acoustic oscillations in a brush discharge corona, for this nonlinear quivering transducer[75] may interact with coherent energetic modes in the zero-point sea.

ACKNOWLEDGEMENT

The author wishes to thank David Faust for helpful discussions.

REFERENCES

ZERO-POINT ENERGY ACTIVITY

1. T. H. Boyer, "Random Electrodynamics: The Theory of Classical Electrodynamics with Classical Electromagnetic Zero-Point Radiation," *Phys. Rev.* D 11 (4), 790 (1975).
2. C. Lanczos, "Matter Waves and Electricity," *Phys. Rev.* 61, 713 (1942).

3. E. G. Harris, *A Pedestrian Approach to Quantum Field Theory*. Wiley Interscience, NY, ch. 10, 1972.
4. C. W. Misner, K.S. Thorne and J.A. Wheeler, *Gravitation*, W.H. Freeman and Co., NY, ch 43-44, 1970.
5. A. Woodcock and M. Davis, *Catastrophe Theory*. Avon Books, NY, 1980.

UNIFORM ACCELERATION OF CHARGE

6. F. Rolrhich, "The Definition of Electromagnetic Radiation," *Il Nuovo Cimento* XXI, 811 (1961).
7. H. A. Atwater, "Radiation From a Uniformly Accelerated Charge," *Am. J. Phys.*, 38 (12), 1447 (1970).
8. V. L. Ginzburg, "Radiation and Radiation Friction Force in Uniformly Accelerated Motion of a Charge," *Sov. Phys. Uspekhi* 12 (4), 565 (1970).
9. D. G. Boulware, "Radiation From a Uniformly Accelerated Charge," *Ann. Phys.* 124, 169 (1980).
10. C. M. Dewitt and W. G. Wesley, "Quantum Falling Charges," *Gen. Rel. & Grav.* 2 (3), 235 (1971).
11. B. S. Dewitt and R. W. Breme, "Radiation Damping in a Gravitational Field," *Ann. Phys.* 9, 220 (1960).
12. D. W. Sciama and P. Candelas, "Quantum Field Theory, Horizons and Thermodynamics," *Adv. Phys.* 30 (3), 327 (1981).
13. T. Fulton and F. Rolrlich, "Classical Radiation From a Uniformly Accelerated Charge." *Ann. Phys.* 9, 499 (1960).
14. W. Pauli, Translated in *Theory of Relativity*, Pergamon Press, NY, p. 93, 1958.
15. R. Vasudevan, "Does a Uniformly Accelerated Charge Radiate?" *Lett. Al Nuovo Cimento* V (6), 225 (1971).
16. A. M. Luiz, "Does a Uniformly Accelerated Charge Radiate?" *Lett. Al Nuovo Cimento,* IV (7), 313 (1970).
17. C. Leibovitz and A. Peres, "Energy Balance of Uniformly Accelerated Charge," *Ann. Phys.* 25, 400 (1963).
18. H. B. Callen and T. A. Welton, "Irreversibility and Generalized Noise," *Phys. Rev.* 83 (1), 34 (1951).
19. P. Candelas and D.W. Sciama, "Is There a Quantum Equivalence Principle?" *Phys. Rev.* D 27 (8), 1715 (1983)
20. B. S. Dewitt, "Quantum Gravity," *Sci. Amr.*, 112 (Dec 1983).

VACUUM POLARIZATION

21. B. Toben, J. Sarfatti and F. Wolf, *Space-Time and Beyond*. E. P. Dutton and Co., NY. pp. 52-53, 1975.

22. I. R. Senitzky, "Radiation-Reaction and Vacuum Field Effects in Heisenberg-Picture Quantum Electrodynamics," *Phys. Rev. Lett.* 31 (15), 955 (1973).

23. F. Scheck, *Leptons, Hadrons and Nuclei*. North Holland Physics Publ., NY, pp. 212-223, 1983.

24. W. Greiner, "Dynamical Properties of Heavy-Ion Reactions - Overview of the Field," *S. Afr. J. Phys.* 1 (3-4), 75 (1978).

25. J. Reinhardt, B. Muller and W. Greiner, "Quantum Electrodynamics of Strong Fields in Heavy Ion Collisions," *Prog. Part. and Nucl. Phys.* 4, 503 (1980).

26. J. S. Greenberg and W. Greiner, "Search for the Sparking of the Vacuum," *Physics Today*, 24 (Aug 1982).

27. G. M. Graham and D.G. Lahoz, "Observation of Static Electromagnetic Angular Momentum in Vacuo," *Nature* 285, 154 (May 1980).

28. F. Roesel, D. Trautmann and R.D. Viollier, "Vacuum Polarization Potential for Two Extended Charge Distributions," *Nucl. Phys.* A 292 (3), 523 (1977).

29. G. Soff, J. Reinhardt, B. Muller and W. Greiner, "Shakeoff of the Vacuum Polarization in Quasimolecular Collisions of Very Heavy Ions," *Phys. Rev. Lett.* 38 (11), 592 (1972).

30. E. A. Rausher, "Electron Interactions and Quantum Plasma Physics," *J. Plasma Phys.* 2 (4), 517 (1968).

31. R. A. Cover and G. Kalman, "Longitudinal, Massive Photon in an External Magnetic Field," *Phys. Rev. Lett.* 33, 1113 (1974).

32. D. B. Melrose and R. J. Stoneham, "Vacuum Polarization and Photon Propagation in a Magnetic Field," *Il Nuovo Cimento* 32 A (4), 435 (1976).

SELF-ORGANIZING SYSTEMS

33. G. Nicolis and I. Prigogine, *Self-Organization in Nonequilibrium Systems*, Wiley, NY, 1977.

34. H. Haken, *Synergetics*, Springer-Verlag, NY, 1971.

ION-ACOUSTIC OSCILLATIONS

35. V. Yu. Bychenkov, A. M. Natonzon, and V. P. Silin, "Anomalous Absorption of Radiation on Ion-Acoustic Fluctuations," *Sov. J. Plasma Phys.* 9 (3), 293 (1983).
36. A. I. Anisimov, N. I. Vinogradov and B. P. Poloskin, "Anomalous Microwave Absorption at the Upper Hybrid Frequency," *Sov. Phys. Tech. Phys.* 18 (4), 459 (1973).
37. M. Waki, T. Yamanaka, H. B. Kang and C. Yamanaka, "Properties of Plasma Produced by High Power Laser," *Jap. J. Appl. Phys.* 11 (3), 420 (1972).
38. Yu. G. Kalinin, D. N. Lin, L. I. Rudakov, V. D. Ryutor and V. A. Skoryupin, "Observation of Plasma Noise During Turbulent Heating," *Sov. Phys. Dokl.* 14 (11), 1074 (1970).
39. H. Iguchi, "Initial State of Turbulent Heating of Plasmas," *J. Phys. Soc. Jpn.* 45 (4), 1364 (1978).
40. E. K. Zavoiskii, et al., "Advances in Research on Turbulent Heating of a Plasma," Proceedings of 4th Conference on Plasma Physics and Controlled Nuclear Fusion Research, pp. 3-24, 1971.
41. A. Hirose, "Fluctuation Measurements in a Toroidal Turbulent Heating Device," *Phys. Can.* 29 (24), 14 (1973).
42. V. Hart, private communication, 1982.
43. Y. Kiwamoto, H. Kuwahara and H. Tanaca, "Anomalous Resistivity of a Turbulent Plasma in a Strong Electric Field," *J. Plasma, Phys.* 21 (3), 475 (1979).
44. M. J. Houghton, "Electron Runaway in Turbulent Astrophysical Plasmas," *Planet. and Space Sci.* 23 (3), 409 (1975).
45. J. D. Sethian, D A. Hammer and C. B. Whaston, "Anomalous Electron-Ion Energy Transfer in a Relativistic-Electron-Beam-Heated Plasma," *Phys. Rev. Lett.* 40 (7), 451 (1978).
46. S. Robertson, A. Fisher and C. W. Roberson, "Electron Beam Heating of a Mirror Confined Plasma," *Phys. Fluids,* 32 (2), 318 (1980).
47. M. Porkolab, V. Arunasalam, and B. Grek, "Parametric Instabilities and Anomalous Absorption and Heating in Magnetoplasmas," International Congress on Waves and Instabilities in Plasmas, Inst. Theoret. Physics, Innsbruck, Austria, 1973.
48. M. Tanaka and Y. Kawai, "Electron Heating by Ion Acoustic Turbulence in Plasmas," *J. Phys. Soc. Jpn.* 47 (1), 294 (1979).

49. Y. Kawai and M. Guyot, "Observation of Anomalous Resistivity Caused by Ion Acoustic Turbulence." *Phys. Rev. Lett.* 39 (18), 1141 (1977).

50. P. J. Baum and A. Bratenahl, "Spectrum of Turbulence at a Magnetic Neutral Point," *Phys Fluids* 17 (6), 1232 (1974).

51. M. Porkolab, "Parametric Instabilities and Anomalous Absorption and Heating of Plasmas," Symposium on Plasma Heating and Injection, Editrice Compositori, Bolona, Italy, pp. 46-53, 1972.

SWEEPING EMISSIONS, WHISTLERS

52. C. D. Reeve and R.W. Boswell, "Parametric Decay of Whistlers—A Possible Source of Precursors," *Geophys. Res. Lett.* 3 (7), 405 (1976).

53. M. S. Sodha, T. Singh, D. P. Singh and R. P. Sharma, "Excitation of an Ion-Acoustic Wave by Two Whistlers in a Collisionless Magnetoplasma," *J. Plasma Phys.* 25 (2), 255 (1981).

54. P. K. Shukla, "Emission of Low-Frequency Ion-Acoustic Perturbations in the Presence of Stationary Whistler Turbulence," *J. Geophys. Res.* 82 (7), 1285 (1977).

55. M. Watanabe, "On the Whistler Wave Solitons," *J. Phys. Soc. Jpn.* 45 (1), 260 (1978).

56. W. C. Gerson and W. H. Gossard, "Sweeping Emissions," *Phys. Can.* 27 (4), 39 (1971).

57. S .R. P. Nayar and P. Revathy, "Anomalous Resistivity in the Geomagnetic Tail Region," *Planet. and Space Sci.*, 26 (11), 1033 (1978).

58. W. L. Webb, *Geoelectricity*, U. of Texas, El Paso, pp. 9-11, 1980.

59. I. M. Podgornyi and R. Z. Sagdeev, "Physics of Interplanetary Plasma and Laboratory Experiments," Sov. Phys. Uspekhi 98, 445 (1970).

INVENTIONS

60. T. H. Moray and J. E. Moray, *The Sea of Energy*. Cosray Research Institute, Salt Lake 1978.

61. T. H. Moray, "Electrotheurapeutic Apparatus," US Patent No. 2,460,707 (1949); contains corona discharge tubes.

62. M. B. King, "Stepping Down High Frequency Energy," Proceedings of the First International Symposium on Nonconventional Energy Technology, University of Toronto, pp. 145-158, 1981.

63. N. Tesla, *Colorado Springs Notes* 1899-1900, Nolit, Beograd, Yugoslavia, 1978.

64. N. Tesla, "Electrostatic Generators," *Sci. Amr.*, 132 (March 1934).

CORONA DISCHARGE, BALL LIGHTNING

65. W. W. Eidson, D. L. Faust, H .J. Kyler, J. O. Pehek and G. K. Poock, "Kirlian Photography: Myth, Fact and Applications," IEEE Electro 78 Convention Proceedings, Special Evening Session, Part One, SS/3, Boston, Mass., pp.1-21, May 23-25, 1978.

66. D. G. Boyers and W. A. Tiller, "Corona Discharge Photography," *J. Appl. Phys.* 44 (7), 3102 (1973).

67. D. Faust, private communication, 1984.

68. J. Corum, "Theoretical Explanation of the Colorado Springs Experiment," The Tesla Centennial Symposium, Colorado College, Colorado Springs (Aug 10-12, 1974).

69. J. D. Barry, *Ball Lightning and Bead Lightning*, Plenum Press, NY, 1980.

70. S. Singer, *The Nature of Ball Lightning*, Plenum Press, NY, 1971.

71. H. W. Secor, "The Tesla High Frequency Oscillator," *Electrical Experimenter* 3, 615 (1916).

72. W. H. Bostick, "Experimental Study of Plasmoids," *Phys. Rev.* 106 (3), 404 (1957).

73. D. R. Wells, "Dynamic Stability of Closed Plasma Configurations," *J. Plasma Phys.* 4 (4), 654 (1970).

74. P. O. Johnson, "Ball Lightning and Self-Containing Electromagnetic Fields," *Am. J. Phys.* 33, 119 (1965).

75. N. Tesla, *Lectures, Patents and Articles*, Nicola Tesla Museum, Beograd, Yugoslavia, pp. L65-68, 1956. (Reprinted by Health Research, Mokelumne Hill, CA 1973.)

COHERING THE ZERO-POINT ENERGY

August 1986

ABSTRACT

By merging theories of the zero-point energy with the theories of system self-organization, it becomes theoretically possible to cohere the zero-point energy as a source. The key component for interacting with the zero-point energy is nuclei of ions in a plasma or electrolyte. A hyperspatial flux model of the zero-point energy is introduced, and it is shown to be affected by abruptly pulsed, opposing electromagnetic fields. This leads to the concept of a "scalar wave" which is modeled as a hyperspatial structure consisting of vortex rings of electric flux. The projection of this structure through three dimensional space yields scalar and longitudinal electromagnetic components. No energy propagates through these components, but energy may propagate through the parallel hyperspace. Experiments are suggested to test the proposed model.

INTRODUCTION

Within our physics today exist theoretical constructs that

may allow the possibility of tapping energy directly out of the fabric of space, creating artificial gravitational fields for stressless, inertialess propulsion, altering the pace of time in a region of space, and even teleportation beyond our space-time continuum. These possibilities arise by combining two branches of theoretical physics: Theories of the zero-point energy (ZPE)[9-17] with theories of system self-organization.[21-26] The theories of system self-organization are the most novel. In 1977 Ilya Prigogine won the Nobel Prize in chemistry for identifying under what conditions a system may evolve from a chaotic, turbulent state to an organized state. The system must be nonlinear, far from equilibrium and have an energy flux through it. The ZPE exhibits these properties. It is highly nonlinear in its interaction with matter or charged particles, it can be driven from equilibrium by abrupt motions of matter or plasma, and it may be a manifestation of an energy flux from a physical hyperspace. The ZPE may fulfill the conditions for self-organization.

By asking the question, "from where does the ZPE come, or from where does a charged particle's electric flux come?", the existence of a physical hyperspace is introduced. This concept is not new to physics. Wheeler[49] shows how it arises by applying general relativity to theories of the ZPE, and Everett[50] shows how it arises in his many worlds interpretation of quantum mechanics. The point of view will be introduced that all matter, elementary particles and fields are physically hyperspatial in nature and that we, like "flat-landers," can only perceive a three dimensional projection of the hyperspatial object or field. From this, the notion of a "scalar wave" or a longitudinal vacuum polarization structure arising from abruptly bucking electromagnetic fields will be introduced and shown to have a hyperspatial form that manifests its 3-space projection with scalar and longitudinal components. It will be shown why the conduction band electrons in macroscopic objects like wires or antennas cannot readily detect this form, but ions or nuclei may be able to interact with it. Finally some frugal, "generic" experiments will be suggested to see if the above conjectures can be supported.

ZERO-POINT ENERGY

What is the zero-point energy? The zero-point energy is the ether, the all-pervading energy that fills the fabric of space. Pre-twentieth century physics viewed space as filled with a material substance that would support the propagation of light. After the Michelson-Morley experiment failed to detect the ether wind, the notion of a material ether was dropped by the physics community. In the 1930's, physicists recognized a term arising from the equations of quantum mechanics and gave it the name zero-point energy. "Zero-point" refers to zero degrees Kelvin and means the energy fluctuations are not thermal in nature. The development of quantum electrodynamics recognized this energy existed in a pure vacuum, and Dirac predicted how electron-positron pair production could arise from it. Boyer recently introduced the viewpoint that quantum mechanical effects arise because of matter's interaction with the zero-point energy[11] and derived its spectral energy density from a postulate of Lorentz invariance.[9] Quantum gravity theories[16,17] show that the ZPE spectrum is altered by gravitational fields or acceleration, and that the curvature of space-time is intimately linked to its action. Quantum electrodynamics shows all particles are intertwined in a vacuum polarization interaction with the ZPE and shows how the interaction yields the mass of an elementary particle.[15] Nonlinear quantum mechanics also yields the mass of an elementary particle through a persistent self-interaction with the zero-point energy while avoiding the renormalization problems of perturbation analysis.[14] A modern view is that elementary particles are a coherence in the zero-point energy[51] and this view can be supported by system self-organization theories.

From where does the zero-point energy come? Recent experiments have shown that the brightness of the zero-point energy is independent of the existence of reflectors and absorbers.[30] This shows that the ZPE does not arise from an electromagnetic propagation in our three dimensional space. Wheeler's geometrodynamics[49] answers the question while at the same time resolving the philosophical problem of the ZPE's infinite energy density arising in quantum mechanics. By applying the formalism of general relativity to the ZPE, Wheeler derives the modern view of the fabric of space. Since energy (or mass) can curve space-time, a sufficiently large energy density pinches off the fabric of space (like a black hole) in the direction of a hyperspace orthogonal to our three dimensional space. The ZPE arises from an electric flux that flows orthogonal to our 3-space (Figure 1). As the flux enters, it manifests itself as a virtual mini positive particle; it leaves through a corresponding virtual mini negative particle. The scale of these particles are on the order of Planck's length, 10^{-33}cm.[49] The size of the electron is on the order of 10^{-13} cm. That is twenty orders of magnitude difference. As the flux passes through our 3-space, there is jitter, and the separation (or pair production) of these mini particles gives rise to a turbulent virtual plasma, sometimes called the "quantum foam."[49] A bias in the direction of this separation induced from charged matter or elementary particles is called "vacuum polarization." The elementary particles themselves may be viewed as vortex ring structures[4]

Figure 1 THE ZERO-POINT ENERGY MAY ARISE FROM AN ORTHOGONAL ELECTRIC FLUX FROM THE FOURTH DIMENSION

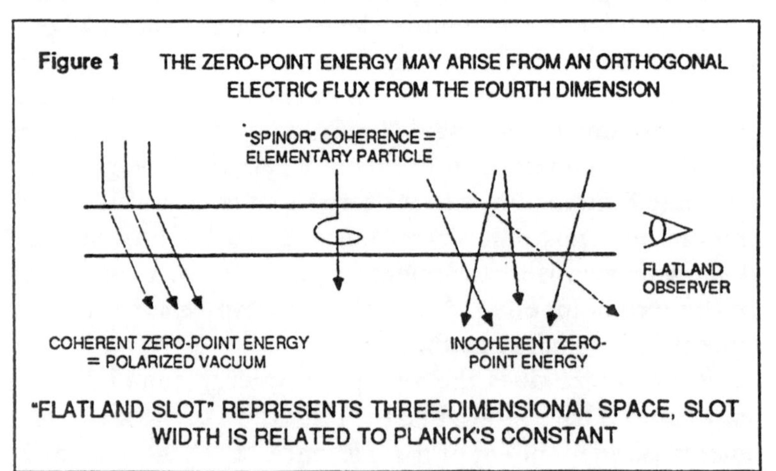

"FLATLAND SLOT" REPRESENTS THREE-DIMENSIONAL SPACE, SLOT WIDTH IS RELATED TO PLANCK'S CONSTANT

maintained by the orthogonal ZPE flux. The turbulent, virtual plasma model of the ZPE is highly nonlinear, highly interactive with matter and is maintained by a flux of electrical energy flowing orthogonally through our 3-space. This offers opportunities for self-organization.

ION OSCILLATIONS

In order to cohere the zero-point energy, one must use those particles or configurations that maximize their interaction with it. Quantum electrodynamics shows that the different elementary particles have different vacuum polarization interactions with the ZPE. In a first order model, nuclei exhibit stable lines of polarization converging toward their centers. On the other hand, conduction band electrons behave as a smeared charged cloud with polarization lines jittering randomly, thus resulting in an equilibrium condition in its interaction with the ZPE. The nucleus' stable lines of vacuum polarization may allow it to launch and detect vacuum polarization displacement currents with which the smeared electron cloud could not readily interact. In addition, the high energy density of the steep vacuum polarization convergence near the nucleus would result in a space-time curvature that would offer stable vacuum polarization displacement channels for the hyperspatial ZPE flux. The nucleus of an ion thus becomes a key component for cohering the ZPE.[8]

Coherently oscillating a large number of ions together may result in a synergistic interaction with the ZPE. The coherent oscillation of ions in a plasma is known as the ion-acoustic mode and, in experiments, it has produced large radiant energy absorption, high frequency spikes, runaway electrons, anomalous plasma heating and anomalous plasma resistance.[8] Moray[52] may have been the first to recognize its importance for interacting with coherent macroscopic vacuum polarization modes (displacement currents) in the ZPE. Moray carefully constructed corona discharge tubes to tune the ion-acoustic oscillations to specific modes in order to step down high frequency energy.[54] Maintaining an ion-acoustic resonance in air is more difficult. Since the plasma frequency is proportional to the ion density, and since more ionization will occur as the energy increases in the resonance, the ion-acoustic plasma frequency will shift up the spectrum as it is excited. Likewise, as it decays, it will emit whistler or sweeper waveforms. The ideal exciter for the ion-acoustic mode in air could be a waveform which rapidly shifts or "chirps" its frequency.[53]

Ion motions or oscillations can be induced in electrolytic solutions as well. Graneau,[38] in experimenting with explosive discharges in salt water, observed a threshold phenomenon related to the sharpness of the pulse excitation. Unless the pulse rise time was sufficiently large, an ordinary discharge would occur in the water and yield no motion. But with the same total pulse energy, when the rise time exceeded a certain threshold, the water would jerk upward explosively. Clearly there is ion motion in this event. Perhaps a similar phenomena happens for those inventors who claim energy anomalies when pulse charging an electrolytic battery. Puharich[39] claimed observing an energy anomaly by electrolysis of water where the excitation signal matched the resonant frequencies of the water molecule's bonds. This would also induce ion oscillations. If abrupt electrolytic ion motion produces a self-organizing interaction with the ZPE (or an orthorotation of its flux), it could explain the energy anomalies claimed in systems where electrolytic ions are excited.

PULSED FIELD OPPOSITION

Opposing electromagnetic fields may produce a direct interaction with the orthogonal ZPE flux. When electromagnetic fields oppose, there is no net field vector, yet the stress on the fabric of space (or the stress-energy tensor of general relativity) increases. It could be described as a region of increased electromagnetic potential. The interior of a Faraday cage is a simple example of opposing fields. The electrons arrange themselves on the skin of the cage such that all field vectors inside are in perfect opposition. Increasing the charge on a Faraday cage is equivalent to compressing a charged balloon symmetrically around the region. The interior field vectors increase in magnitude but remain in opposition. Quantum gravity theories show that a high potential or large components of the stress energy tensor can alter the action of the ZPE. DeBroglie[18,19] suggests that since a particle interacts with a "hidden thermodynamics" (ZPE), if the action of this changes, the particle's proper mass can change. Nonlinear quantum mechanics predicts a similar mass ZPE coupling. This suggests an experiment where a system whose characteristic frequency or behavior is a function of the electron's proper mass could be affected through a charged Faraday cage.

If abrupt bucking fields were impressed around a nucleus or ion lattice, a direct orthorotation of the ZPE flux might be accomplished (Figures 2 and 3). The leading edge of the pulsed fields pinch the orthogonal flux and build pressure; the sudden release allows the energy to snap-back into our 3-space if there are stable vacuum polarization channels for it to follow. Nuclei could provide these channels as well as provide a continuous space-time metric curvature toward the orthogonal hyperspace due to its large mass-energy density.

A frugal experiment to test this idea involves a caduceus coil.[44] A caduceus coil has two perfectly symmetrical wind-

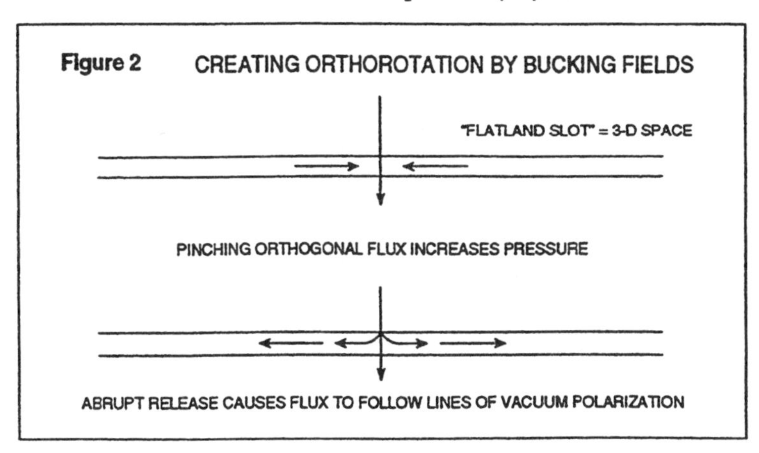

Figure 2 CREATING ORTHOROTATION BY BUCKING FIELDS

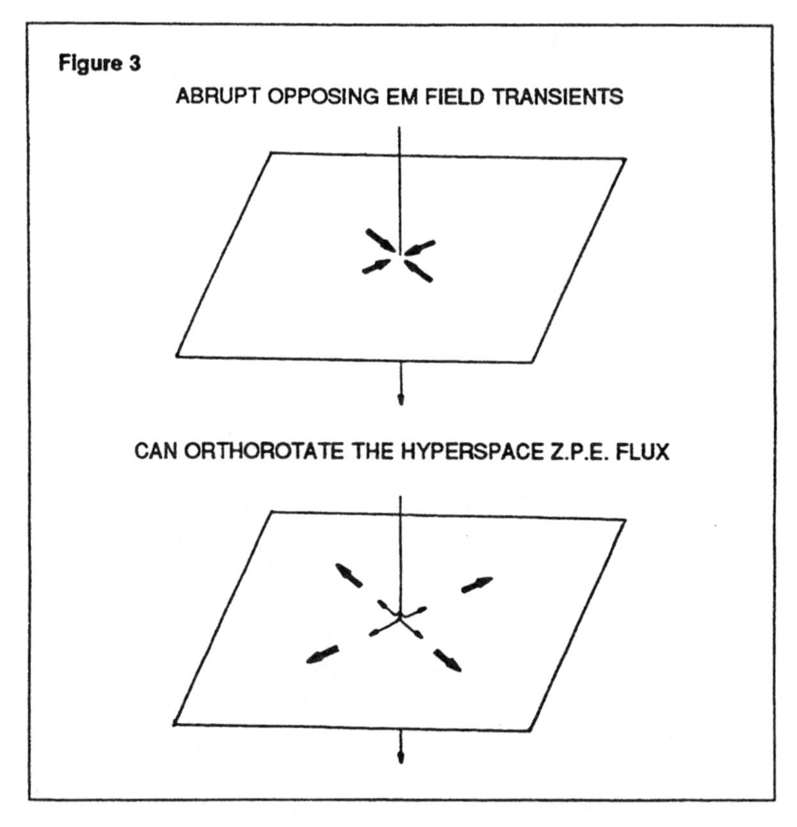

Figure 3

ings on a ferrite core of opposite helicity. When pulsed, the nuclei of the ferrite lattice experiences the abrupt bucking magnetic fields. Smith claimed a "tempic field" would be produced in such an event which would alter the pace of time for objects in it.[43] General relativity can support this idea if sufficient ZPE is orthorotated such that a space-time curvature occurs. A simple experiment would be to spike pulse the caduceus coil with spikes at a rapid repetition rate and then affect the characteristic frequency of some system (e.g. an oscillator) near the coil. Repeat the experiment with the oscillator inside a Faraday shield and the coil outside. Instead of the oscillator, another detector could be a corona. One would then observe the corona's activity in response to the activated caduceus coil. Pre-exciting the corona at its ion-acoustic resonance may produce a synergistic surprise. If the ZPE flux is orthorotated by the abrupt bucking fields of the caduceus coil and sufficient energy were orthorotated, then the bending of the fabric of space-time could occur. If the bending became extreme, teleportation could result.[45]

VORTEX RINGS

It is well recognized in hydrodynamics that abruptly pulsing a turbulent medium can give rise to vortex ring formations. Theories of system self-organization recognize that vortices and vortex rings are soliton forms that can spontaneously arise out of turbulence if a trigger can be provided that drives the system far from equilibrium.[1,2] The vortex or vortex ring is an archetype form that appears at all levels in nature from galaxies to elementary particles. The vortex ring has been experimentally observed forming in plasma and is called a "plasmoid."[3] It has been used to model ball lightning,[5] elementary particles,[4] and even a photon.[6] By assuming a hyperspatial flux model for the ZPE turbulence, an application of self-organizing principles would predict that abruptly pulsed opposing electromagnetic fields would give rise to a hyperspatial vortex ring structure that is symmetrically located (or launched) around the region of the abruptly bucking fields. The vortex ring energy is circulating in the hyperspace on both sides of our 3-space "flat-land slot." Two pairs of counter-rotating vortex rings are produced: One on the rising edge of the bucking field pulse and one on the trailing edge. The vortex energy content increases dramatically with the rise time of the bucking pulse.[55] An experiment to demonstrate this involves winding a caduceus coil with symmetric perfection, for it is the pulse *edges* that must align while propagating through each of the opposing windings. The vortex ring model matches Smith's description[43] of the "tempic field" toroidal forms activated by the caduceus coil.

SCALAR WAVES

The hyperspatial vortex rings can be applied to model the effects made by what has popularly come to be known as "scalar waves," "longitudinal waves," "columnar standing waves," or "non-Hertzian Tesla waves." Figure 4 illustrates a "flat-land" analogous view of the hyperspatial vortices. Every possible pair of horizontally and vertically adjacent flow circles is a vortex ring intersecting through the paper. Note that in the "flat-land slot" the situation appears static. In the regions where the circular flux orthogonally enters the "slot," a region of positive potential appears; likewise, there is a region of negative potential where flux orthogonally leaves. In between these regions there are longitudinal (vacuum polarization displacement) components that alternate from region to region. The dynamics of the hyperspatial vortices project a static vacuum polarization structure in the flat-land slot. *No energy propagates in the flat-land slot.* (This fulfills the conditions of gauge theories where both the scalar and longitudinal components are allowed as long as there is no net 3-space energy propagation.[34]) However, there is energy propagation in the hyperspace parallel to the flat-land slot guided by the horizontal vortex rings. The phrase "scalar wave" now has the following meaning: The word "scalar" reflects the three dimensional projection of a perceived static vacuum polarization, and the word "wave" refers to the guided propagation of energy through hyperspace in a direction parallel and adjacent to our 3-space.

An experiment to show that energy could be guided through the parallel hyperspace involves guiding energy through thin wires to light a load bank of bulbs without heat-

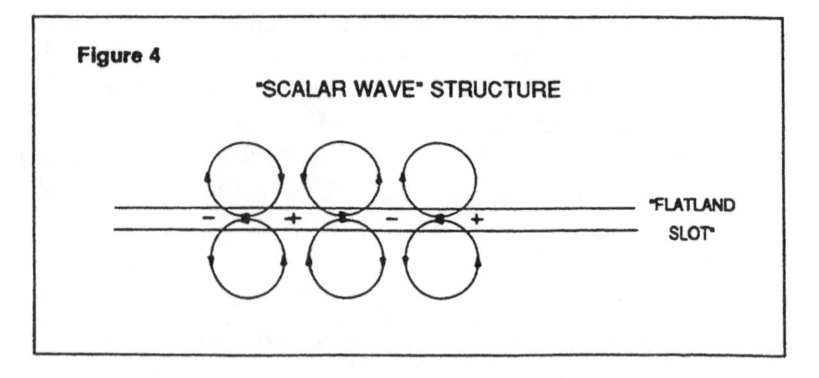

ing the wire. Figure 4 can be used to illustrate why conduction electrons would not respond to the scalar wave. The wire is represented by the "flat-land slot." The vertical vortex rings are threaded by the wire and the horizontal vortex rings guide the energy. The conduction band electrons are a smeared charge cloud along the entire length of the wire and will simply settle into a static polarized condition along the wire. (Note that adjacent longitudinal components always buck, resulting in no net conduction electron current flow.) However, if there were free ions in the vicinity of the rotating vortex flow, they could follow the flow coherently. It is the ions in the light bulb plasma that respond to the scalar wave. One could also detect the scalar wave with the ions of a corona. In addition, the corona may form into a vortex ring plasmoid. If the plasmoid itself orthorotated the ZPE flux, it could persist (like ball lightning) even after the original excitation was stopped. The guiding of power along thin wires without heating, or the persistence of a plasmoid after the input power is stopped, are two spectacular experiments that could demonstrate the hyperspatial physics.

A third experiment involves penetration of electromagnetic (Faraday) shielding. The conduction electrons on the shield would only respond to the static 3-space projection of the scalar wave. But the hyperspatial components (horizontal vortex rings of Figure 4) could guide the energy parallel to our 3-space through the hyperspace channels where it could be detected by the ions in a corona inside the shielded region. Shielding penetration would also be a convincing experiment of hyperspatial electromagnetics.

Perhaps the largest effects can be achieved by subjecting ion vortices, plasma vortex rings, or vortex ring segments directly to abruptly pulsed opposing magnetic fields. Such configurations exist in the air gap of Gray's motor,[40] along the rotor of Searl's or Carr's flying disks,[41] and in the Swiss ML convertor.[56] Claims of "free energy" and/or gravitation anomalies are associated with these inventions.

CONCLUSION

By modeling the ZPE as an energy flux from a physical hyperspace orthogonal to our 3-space, and showing that this flux is coherently interactive with nuclei through their stable vacuum polarization channels, the possibility of a ZPE coherence or orthorotation arises from the principles of system self-organization theories. The use of abruptly pulsed opposing electromagnetic fields was identified as the trigger to induce the orthorotation of the ZPE flux, and it could result in a hyperspatial vortex ring structure that was given the name "scalar waves". This form exhibits scalar or static components in its 3-space projection, yet it guides energy through a hyperspace channel parallel to our 3-space. It was shown that the nuclei of the ions could detect this scalar wave but that conduction band electrons would not. Experiments involving guiding power along thin wires without heating, or Faraday shielding penetration were suggested as experiments to demonstrate the hyperspatial physics. The combination of ion vortices or vortex rings with abruptly pulsed bucking magnetic fields could result in the direct manifestation of energy and gravitational anomalies. It is hoped that many inventors, by working with these ideas, would freely donate a frugal, "generic" experiment that could be readily repeated. For it is the repetition of an experiment that will spread across our planetary consciousness like a wave and usher in the forthcoming golden age.

ACKNOWLEDGEMENTS

The author wishes to thank those who freely donated their support and talents in order to help clarify this presentation.

COHERING THE ZERO-POINT ENERGY ANNOTATED BIBLIOGRAPHY

August 1986

In order to theoretically allow the coherence of the zero-point energy (ZPE), two fields of modern physics must combine: theories of the zero-point energy interacting with matter, and theories of system self-organization. The latter theories identify the conditions for self-induced coherence: The system must be nonlinear, far from equilibrium, and have an energy flux through it. The dynamics of the ZPE and its interaction with matter are nonlinear. It can be driven from equilibrium with abrupt motions of matter (e.g., electric discharges), and it may be a manifestation of an electrical flux that flows orthogonally through our three-dimensional space from a hyperspace. The dynamics of the ZPE can fulfill the conditions for self-organization. The following references are presented in support of this conjecture.

[The author's speculations are enclosed in brackets.]

1. S. V. Antipov, M. V. Nezlin, A. S. Trubnikov, "Rossby Autosoliton," *JETP Lett.* 41 (1), 30-33, (1985).

   The most persistent and stable soliton vortex has been created in the laboratory. It self-organizes from oppositely-directed streams and is thought of as the model for the great red spot of Jupiter. [The stability of this form may make it a candidate for the appropriate model of elementary particles self-organizing from a ZPE hyperspatial electric flux.]

2. A. V. Panifilov, A. T. Winfree, "Dynamical Simulations of Twisted Scroll Rings in Three-Dimensional Excitable Media," *Physica* 17D, 323-330, (1985).

   The self-organizing formation of scroll rings (convergent vortex filaments that close into rings) has been observed experimentally in chemical and biological systems and now has been numerically simulated in three dimensions. The spiral flow around the filament does not radiate from the vortex but rather converges onto it. Scroll ring pair creation is expected. [The details of the filament formation should lend insight to the vortex experiments of Schauberger (Reference 42), ball lightning, plasmoid and perhaps elementary particle formation. The vortex ring is the archtype form for self-organization in the ZPE.]

3. W. H. Bostick, "Experimental Study of Plasmoids," *Phys. Rev.* 106(3), 404, (1957).
   An experiment is reported where vortex ring structures occur in a plasma. Plasmoid pair production is observed. A "quantum condition" in the ratio of toroidal to poloidal diameters is identified as needed for stability.
4. W. H. Bostick, "The Gravitationally-Stabilized Hydromagnetic Model of the Elementary Particle," Gravity Research Foundation, New Boston, N.H., (1961).
   A vortex ring structure for the electron is proposed.
5. P. O. Johnson, "Ball Lightning and Self-Containing Electromagnetic fields," *Am. J. Phys.* 33, 119, (1965).
   A vortex ring structure for ball lightning is presented.
6. W. M. Honig, "A Minimum Photon Rest Mass Using Planck's Constant and Discontinuous Electromagnetic Waves," *Found Phys.* 4(3), 367-80, (1974).
   A vortex ring model of the photon, sustained by a two fluid ZPE model, expands at the speed of light and has implications for nonlocal connectivity.
7. S. Iizuka, H. Janaca, "Nonlinear Evolution of Double Layers and Electron Vortices in an Unstable Plasma Diode," *J. Plasma Phys.* 33(1), 29-41, (1985).
   Nonlinear evolution of large-amplitude relaxation oscillation in a unstable plasma diode is numerically simulated as well as experimentally observed. It is related to plasma ion-acoustic activity. The plasma two-stream instability gives rise to a vortex formation which causes positive potential spikes. [This is an excellent reference for understanding the radiant energy device of T. H. Moray. The plasma diode can be related to the Moray valve.]
8. M. B. King, "Macroscopic Vacuum Polarization," Proceedings of the Tesla Centennial Symposium, International Tesla Society, Colorado Springs, 99-107, (1984).
   It is speculated that the ion-acoustic oscillations in a plasma may interact with macroscopic vacuum polarization or organ-

ized modes in the ZPE that conduction electrons do not readily detect. This may be the operating principle behind the various discoveries of T. H. Moray.

9. T. H. Boyer, "Derivation of the Blackbody Radiation Spectrum Without Quantum Assumptions," *Phys. Rev.* 182(5), 1374-83, (1969).
   The author derives the spectrum of the ZPE by assuming Lorentz invariance. In order for all inertial observers to measure the same ZPE spectrum, the energy density must increase with increasing frequency without a cutoff. This is the source of diverging energy calculations in quantum mechanics.
10. T. H. Boyer, "Thermal Effects of Acceleration Through Random Classical Radiation," *Phys. Rev.* D 21(9), 2137-48 (1980).
    Boyer shows that a uniformly accelerating observer would see the ZPE spectrum as a thermal spectrum. This classical theory begets the same result as derived from quantum field theory.
11. T. H. Boyer, "Random Electrodynamics: The Theory of Classical Electrodynamics With Classical Electromagnetic Zero-Point Radiation," *Phys. Rev.* D 11(4), 790-808, (1975).
    Boyer shows that quantum effects arise because of matter's interaction with the zero-point energy.
12. S. J. Putterman, P. H. Roberts, "Random Waves in a Classical Nonlinear Grassman Field," *Physica* 131 A, 51-63 (1985).
    Fermi statistics arise from the particles' nonlinear interaction with the ZPE. The authors recognize that the nonlinear Langevin formalism (see next reference) allows certain nonlinear waves to extract energy from some modes of the zero-point energy. The authors are critical of this result. [This result is a natural occurrence that would arise out of nonlinear, self-organization theories. It is a prediction that awaits an experiment.]
13. E. M. Lifshitz, L. P. Pitaevskii, *Statistical Physics, Part 2,* Pergamon Oxford, (1980); "Hydrodynamic Fluctuations," pp. 369-73.
    This general nonlinear hydrodynamic model is applicable even to quantum mechanical systems interacting with the zero-

point fluctuations. Putterman (previous reference) disagreed with this formalism for it allowed energy to be extracted from the zero-point fluctuations.

14. P. B. Burt, *Quantum Mechanics and Nonlinear Waves,* Harwood Academic, N.Y., (1981).
This is an excellent text which promotes the view that all quantum mechanical particles and systems arise from a persistent, nonlinear self-interaction with the ZPE. The author points to the limitations of perturbation analysis (linear approximations to the nonlinear system), and shows how many of the difficulties due to diverging calculations can be resolved by using the full nonlinear solutions. [The self-organizing nonlinear system (or particle) self-organizes or coheres a certain portion of the infinite ZPE giving rise to a finite observable energy. The nonlinear system creates its own cutoff, thus avoiding divergences that would arise in perturbation analysis.]

15. I. R. Senitzky, "Radiation-Reaction and Vacuum Field Effects in Heisenberg-Picture Quantum Electrodynamics," *Phys. Rev. Lett.* 31(15), 955, (1973).
The author shows that all elementary particles are inseparably intertwined with the zero-point energy, and this interaction is the basis of a charge particle's radiation characteristics.

16. N. D. Birrell, P. C. W. Davies, *Quantum Fields in Curved Space,* Cambridge University Press, NY, (1982).
This text is an overview of quantum gravity theories where the ZPE plays a crucial role. In flat (Minkowski) space-time, the infinite ZPE can be renormalized away. This yields accurate results because of an equilibrium situation. But, in a nonlinear curved space, a far-from-equilibrium fluctuation can result in a big bang.

17. B. S. Dewitt, "Quantum Gravity," *Sci. Am.,* 112 (Dec 1983).
This article introduces quantum gravity theories where the ZPE plays a crucial and sometimes active role.

18. L. DeBroglie, *Nonlinear Wave Mechanics,* Elsevier Pub. Co., NY (1960).
DeBroglie proposes that quantum particles may be modeled by a particle that is guided by a nonlinear pilot wave. The frequency of the particle's "internal clock" and pilot wave always maintain phase coincidence. As a result of the wave-particle's

interaction with a "hidden thermodynamics" [ZPE], the quantum particle exhibits a dynamic with variable proper mass. [DeBroglie's model has many features that are similar to other nonlinear models where the persistent self-interaction with the ZPE results in the observed net mass of the particle].

19. L. DeBroglie, "The Reinterpretation of Wave Mechanics," *Found. Phys.* 1(1), 5-15 (1970).

    DeBroglie presents an overview of his interpretation of quantum mechanics. (See previous reference.) [By linking the proper mass of the particle to the frequency of its quantum wave, the particle can exhibit a variable proper mass. This result would manifest itself as a Bohm-Aharonov effect (see next reference) and would predict a charged quantum particle's mass could be altered even through Faraday shielding, if a sufficient potential were applied to the shield. The mass alters because the action of the ZPE (DeBroglie's "hidden thermodynamics") changes under different potentials. Note quantum gravity theories (references 16, 17) should predict a similar effect since the stress-energy tensor changes under the potential which changes the action of the vacuum fluctuations and these, in turn, exhibit a persistent interaction with the quantum particle.]

20. Y. Aharonov, E. Bohm, "Significance of Electromagnetic Potentials in the Quantum Theory," *Phys. Rev.* 115(3), 485, (1959).

    The authors successfully predict the results of experiments that show that an electron can be affected by a change in potential in the absence of an electromagnetic field (see reference 46).

21. A. Hasegawa, "Self-Organization Processes in Continuous Media," *Adv. Phys.* 34(1), 1-42, (1985).

    This paper reviews the behavior of nonlinear, dissipative continuous media. Such media can exhibit the formation of ordered structures even when starting from an initially turbulent state. Examples discussed include magnetohydrodynamic fluids, magnetized plasmas, and atmospheres of rotating planets. The relationship between the onset of chaos and self-organization in a soliton system as well as vortex solutions is also discussed. [This formalism applied to the ZPE may predict the conditions under which an interactive charge particle system could cohere the ZPE.]

22. M. Suzuki, "Fluctuation and Formation of Macroscopic Order in Nonequilibrium Systems," *Prog. Theor. Phys. Suppl.* 79, 125-140, (1984).
The role of fluctuation and nonlinearity in the formation process of macroscopic order is discussed. A coherent interaction model is also introduced to study self-organizing processes.

23. S. Firrao, "Physical Foundations of Self-Organizing Systems Theory," *Cybernetica* 17(2), 107-24, (1984).
This paper deals with the contradiction between the law of increased entropy and the fundamental hypothesis of any theory of self-organizing systems. The conflict is resolved by a criticism of the law of increased entropy.

24. Yu. L. Klimontovich, M. V. Lomonosov, "Entropy Decrease During Self-Organization and the S Theorem," *Sov. Tech. Phys. Lett.* 9(12), 606-7, (1983).
The authors rigorously prove that the entropy decreases in a self-organization process and call this result the "S Theorem" (S stands for self-organization).

25. H. Haken, *Synergetics,* Springer-Verlag, NY, (1971).
This text identifies, via system theory mathematics, the conditions for self-organization. The formalism can then be applied to any system. [This could include nonlinear theories of the ZPE and its interaction with matter.]

26. I. Prigogine, I. Stengers, *Order Out of Chaos,* Bantam Books, NY, (1984).
This book is a layman's description of Prigogine's Nobel Prize winning contribution to the field of self-organization in thermodynamics.

27. L. de la Pena, A. M. Cetta, "Origin and Nature of the Statistical Properties of Quantum Mechanics," *Hadronic J. Suppl.* 1(2), 413-39, (1985).
A theory of stochastic electrodynamics is presented that shows that quantum stochasticity is due to zero-point radiation.

28. F. Winterberg, "Nonlinear Relativity and the Quantum Ether," *J. Fusion Energy,* 3(2), 7-21, (1985).
This paper presents an heuristic procedure by which the Lorentz transformations follow from the interactions charac-

terized by the quantum mechanical commutation rules. A nonlinear generalization of the Lorentz transformations is derived which departs from special relativity at very high energies and establishes the observable existence of a substratum (ether). This departure from Lorentz' invariance yields a finite zero-point vacuum energy. In a limiting case, special relativity is recovered, but the zero-point energy diverges. The theory satisfies the principle that the space-time structure should be determined from interactions instead of being postulated *a priori*.

29. J. P. Ralston, "Covariant Method for Soliton Matrix Elements," *Phys. Rev.* D 33(2), 496-505, (1986).

    Wave functions for solitons and other collective states are constructed in this covariant nonlinear field theory. A generalized coherent-state expansion is presented which yields the same results as those achieved using standard Hamiltonian quantization procedures. Divergences from zero-point fluctuations are discussed. [This nonlinear theory could predict inducing coherence in the ZPE if the conditions for self-organization were met.]

30. O. H. Abroskina, G. Kh. Kitaeva, A. N. Penin, "The Effective Brightness of Zero-Point Fluctuations of the Electromagnetic Vacuum of Parametric Scattering of Light," *Sov. Phys. Dokl.* 30(1), 67 (1985).

    This paper summarizes an experiment that shows that the effective brightness of the zero-point fluctuations is independent of the presence of reflection and absorption.

    [This implies that the ZPE source is not radiation in our three-dimensional space, but rather must be from an orthogonal hyperspace.]

31. O. Klein, "The Atomicity of Electricity as a Quantum Theory Law," *Nature* 118, 516, (1926).

    As well as supporting Kaluza's five-dimensional unified field theory, Klein suggests that the origin of Planck's quantum may be due to a periodicity in the fifth dimension. The small value of the characteristic length of this periodicity (which is related to Planck's constant) may explain the nonappearance of the fifth dimension in ordinary experiments as the result of averaging over the fifth dimension. [This relates the thickness of the "flatland slot" diagram of three dimensional space to

Planck's constant and a physical hyperspace such as Kaluza's fifth dimension.]

32. S. Hacyan, A. Sarmiento, G. Cocho, F. Soto, "Zero-Point Field in Curved Spaces," *Phys. Rev.* D 32(4), 914-919, (1985).
It is concluded from quantum field theory in curved spaces that the radiation produced by gravitational fields or by acceleration is a manifestation of the zero-point field and of the same nature (whether real or virtual).

33. P. Lorrain, D. Corson, *Electromagnetic Fields and Waves,* W. H. Freeman & Co., San Francisco, (1970).
This text contains an heuristic derivation of the magnetic field as the relativistic transformation of the electric field. For any force a magnetic field analog would be observed regardless of whether it was due to a spring or gravitational field (p. 251). The drift velocity of electrons in a conductor at room temperature is one foot per hour; yet even at this small velocity, a relativistic effect is manifested as the magnetic field (p. 279).

34. I. J. R. Aitchison, *An Informal Introduction to Gauge Field Theories,* Cambridge University Press, (1982).
This monograph reviews the developments in gauge field theories and gives physical interpretations to the mathematical results. The author recognizes that in quantum electrodynamics the scalar component and longitudinal component of the vacuum operators need not be zero, but can be nonzero if their components cancel one another: "The vacuum is not necessarily simply the state with no longitudinal or scalar photons," but, "the energy of the longitudinal and scalar photons is zero (they cancel each other) in allowed states" (p. 44). [This may support the notion of a scalar wave that contains no field energy, yet can affect the DeBroglie wave of a quantum particle. The DeBroglie wave may be viewed as a ZPE flux (from a hyperspace) feeding the particle to create its mass. If this is true, then a potential change or a scalar wave interaction could alter the particle's mass (reference 18).]

A discussion on ferromagnetism noted, "the vacuum state of a quantum field theory is analogous to the ground state of an interacting many body system" (p. 75). Also, "...the ground state of a complicated system (for example, one involving interacting fields) may well have unsuspected properties—which may indeed be very hard to prove as following from the Lagrangian. But, we can postulate (even if we cannot prove) properties of the quantum field theory vacuum which are analogous to those of the ground states of many physically interesting many-body systems—such as ferromagnets, superfluids and superconductors..." (p. 75). It was also noted that ferromagnets could have a nonzero vacuum expectation for both massless and massive excitations (p. 75). [Perhaps this could lend insight as to why the ferrite core is important for successful activation of the caduceus coil (reference 43). It is through the nuclei of the ferrite lattice where the ZPE orthorotated flux would occur in response to the abrupt magnetic opposition from the caduceus windings when the coil is pulsed.]

35. H. Goldstein, *Classical Mechanics,* Addison Wesley, Reading, Mass. (1950).
Hamilton's derivation of a wave equation from purely classical physics matches Schrödinger's equation up to an arbitrary constant. Hamilton could have discovered the Schrödinger equation in 1834 if he had an experimental reason to set this arbitrary constant to Planck's constant, *h* (p. 314).

36. R. Bass, "Self-Sustained Non-Hertzian Longitudinal Wave Oscillations as Rigorous Solutions of Maxwell's Equations for Electromagnetic Radiation," Proceedings of Tesla Symposium, pp. 89-90, International Tesla Society, Colorado Springs, (1984).
The author presents a solution of Maxwell's equations consisting of a helical propagation closing into a torus.

37. T. E. Bearden, "Tesla's Electromagnetics and Its Soviet Weaponization," Proceedings of the Tesla Centennial Symposium, pp. 119-38, International Tesla Society, Colorado Springs, (1984).
In this speculative paper, the author argues that the Soviet Union has developed a full-scale electromagnetic scalar wave technology. A scalar wave is defined as a propagating organization in the ZPE created by abruptly bucking electromagnetic fields. The scalar wave interacts with nuclei due to the nonlinear space-time curvature near their vicinity [as well as to their stable lines of vacuum polarization]. The scalar wave is a ZPE coherence that can curve space-time and result in an artificial

gravitational field. The higher the frequency of pulsed field opposition, the greater the yield from the ZPE. [This may be augmented by subjecting nuclei to pulse field opposition, since the nuclei offer stable lines of vacuum polarization that would act as orthorotation channels to the ZPE flux. The ZPE flux can be viewed as a hyperspatial flow orthogonal to our three dimensional space. Bucking fields pinch this flow and build pressure. Abrupt release of the field opposition then allows a "snap-back" reaction that can orthorotate this flux into our 3-space if stable vacuum polarization channels are available. The high frequency excitation allows an interaction with the more energetic, high-frequency modes of the ZPE. Abruptly pulsing a caduceus coil (reference 43) may provide experimental verification of this conjecture.]

38. P. Graneau, P. N. Graneau, "Electrodynamic Explosions in Liquids," *Appl. Phys. Lett.* 46(5), 468-70, (1985).
An experiment is reported where electric arc currents produce explosions through salt water by electrodynamic forces. The explosive phenomena can be explained with the aid of longitudinal Ampere forces, but not with traditional Lorentz forces. A threshold phenomena is observed below which an ordinary discharge produces no motion in the liquid, but above which a violent motion is induced. [During the explosion, there is obviously ion-acoustic activity. This experimental arrangement may make a convenient transmitter for generating a macroscopic vacuum polarization (reference 8). A similar phenomenon may occur in those "free energy" devices that pulse charge a battery's electrolyte. It may be more convenient for experimental purposes to induce ion-acoustic activity in an electrolytic solution than in a gas discharge plasma tube.]

39. A. Puharich, "Water Decomposition by Means of Alternating Current Electrolysis," Proceedings of the First International Symposium on Nonconventional Energy Technology, Toronto, pp. 49-77, (1981).
The author experimentally decomposed water to hydrogen and oxygen by an electrical excitation at a mixture of frequencies that match the water molecule's resonant frequencies. The author claims a net energy gain; i.e., more energy is produced when the hydrogen and oxygen recombine than it took to separate them. [Where did the energy come from? Perhaps by inducing ion-acoustic resonance in the water, a vacuum-

polarized, self-organizing interaction with the ZPE occurs. This experiment has similarities to "free energy" devices that pulse a battery's electrolyte.]

40. E. V. Gray, "Pulsed Capacitor Discharge Electric Engine," U.S. Patent 3,890,548, (1976).
Gray's motor works on the principle of magnetic repulsion where the magnetic fields are pulsed and opposing. Energizing the rotor magnetic field requires shooting a current pulse across the air gap. The resulting spark plasma in the air gap would then experience the pulsed magnetic fields. [The discharge across the air gap could then form into a helical plasma filament that bends in the direction of the circumferential air flow in the gap due to the rotor motion. This curved plasma filament would resemble a piece of a plasma vortex ring observed in plasmoids and ball lightning. The plasma vortex could be a self-organizing structure that coheres the ZPE. A similar electrical discharge occurs in the "antigravity" disks of Searl and Carr (next reference).]

41. W. P. Baumgartner, *Energy Unlimited,* Issue 20 (1986), P. O. Box 35637, Station D, Albuquerque, NM 87176.
This issue contains numerous articles on the levity disks of J. R. P. Searl and O. T. Carr. These disks are similar and exhibit both anomalous energy production and "antigravitational" behavior. The disks shoot helical electrical discharges radially along their rotors. [The rapid rotation then would bend these discharge filaments into a piece of a vortex ring structure similar to plasmoids and ball lightning. Such a structure could exhibit self-organizing behavior that coheres the ZPE.]

42. B. Frokjaer-Jensen, "The Scandinavian Research Organization on Nonconventional Energy and the Implosion Theory (Viktor Schauberger)," Proceedings of the First International Symposium on Nonconventional Energy Technology, Toronto, pp 78-96, (1981).
This paper overviews the vortex studies of Viktor Schauberger. Energy production was claimed when using an imploding, water-vortex generating apparatus. A bluish corona was observed around the apparatus. [Ions in the water would allow the water-vortex to exhibit aspects of a plasma vortex. The plasma vortex could exhibit self-organizing behavior that coheres the ZPE.]

43. W. B. Smith, *The New Science*, Fern-Graphic Publ., Mississauga, Ontario, (1964).
This esoteric work claims the energized caduceus coil (opposing helical windings on a ferrite core) creates a "tempic field." [This is a region where the time component of the space-time metric is altered. Such a claim might be supported if the ZPE were a hyperspatial flux orthogonal to our three-dimensional space, and the pulsed magnetic field opposition on the ferrite lattice induced an orthorotation of this flux (see references 34 and 37).]

44. G. Burridge, "The Smith Coil," *Psychic Observer*, 35(5), 410-16, (1979).
This article explains how to wind a caduceus coil (previous reference) and reports on some observations made by investigators experimenting with the coil.

45. W. L. Moore, C. Berlitz, *The Philadelphia Experiment: Project Invisibility*, Grosset & Dunlap, NY, (1979).
The authors research persistent rumors that in World War II the U.S. Navy did an experiment where, in attempting to bend light and radar waves around a ship by intense, pulsed magnetic fields, they accidently teleported the ship. [If such an event did occur with technology of the 1940's, it would not have been overly complicated. Curving light around an object requires curving space-time. The only energy strong enough to do this would have to come from orthorotating the ZPE flux. Creating intense, pulsed, opposing magnetic fields with a caduceus coil is a candidate for accomplishing this. If too much ZPE were orthorotated, it would bend space-time too much, and the object would leave our three-dimensional continuum.]

46. S. Olariv, I. I. Popescu, "The Quantum Effects of Electromagnetic Fluxes," *Rev. Mod. Phys.* 57(2), 339-436, (1985).
This is the most extensive review article published to date on the Bohm-Aharonov effect (reference 20). The physical meaning of a localized pure potential (with zero electric and magnetic fields) is debated. [Nonlinear mass changes, quantum gravity, nonlinear persistently self-interacting fields and the ZPE have yet to formally enter into the debate.]

47. S. Seike, *The Principles of Ultrarelativity,* G-research Laboratory, Tokyo, Japan, (1978).

Seike proposes the existence of a physical hyperspace with an electrical energy flux that flows orthogonal to our three dimensional space. To orthorotate this flux into our space requires a four dimensional rotation. Seike calculates how the three dimensional projection of this "hyper" rotation would appear in our 3-space using four dimensional Euclidian geometry. Seike's main theme is that by actualizing the dynamics of the 3-D projection with the motion of charge, the hyperspatial form is created.

48. C. W. Cho, "Tetrahedral Physics", 449 Izumi, Komae City, Tokyo, Japan, (1971).

Cho describes in detail one of Seike's (previous reference) hyperspatial, four-dimensional rotating forms called the "resonating electromagnetic field" (RMF). The form is generated by rapidly switching electric charge in a specific way among four spheres located at the vertices of a tetrahedron. The dynamics of the switching are such that there are two modes of rotation orthogonal to each other: a rotational mode and a precessional, oscillating "inside out" mode. The projected hyperspatial structure is described as a "dynamical Klein bottle." It is predicted that experimentally switching the charges in the described manner will cause the apparatus to exhibit gravitational and inertial anomalies.

49. J. A. Wheeler, *Geometrodynamics,* Academic Press, New York, (1962).

Wheeler derives the modern view of the fabric of space by applying the formalism of general relativity to the ZPE. Due to the large energy density, the fabric of space pinches into blackhole-whitehole pairs that may connect distant regions nonlocally through "wormholes." The fabric of space appears as a turbulent virtual plasma consisting of particles whose size is on the order of Planck's length, 10^{-33} cm. The energy density of electric flux passing through each particle is enormous: 10^{93} g/cm^3. [If even just a slight coherence could be induced in the ZPE, it could yield sufficient energy for all needs.]

50. H. Everett, "The Theory of the Universal Wave Function," in B. S. Dewitt, N. Graham *The Many-Worlds Interpretation of Quantum Mechanics,* pp 3-130, Princeton Uni-

versity Press, (1973).
By allowing everything, including the entire universe, to be a quantum mechanical system, Everett derives an interpretation of quantum mechanics that yields an infinite number of universes "parallel" to our own. This is embedded in a superspace (hyperspace) structure.

51. B. Toben, J. Sarfatti, F. Wolf, *Space-Time and Beyond,* E. P. Dutton, New York (1975).
Toben presents a pictorial, layman's introduction to the modern view of space-time which encompasses *Geometrodynamics* (reference 49) and the *Many-Worlds Interpretation* (reference 50). Sarfatti's speculations on the nature of consciousness are also included.

52. T. H. Moray, J. E. Moray, *The Sea of Energy,* Cosray Research Institute, Salt Lake, (1978).
The history of the discovery of T. H. Moray's radiant energy invention is presented with many testimonials from witnesses. The final invention produced power on the order of 50 kilowatts. In his attempts to achieve lossless resonance, Moray realized electron conduction was too lossey and stressed the importance of ionic oscillations.

53. G. Doczi, *The Power of Limits,* Shambhala, Boston, (1981).
Doczi shows with extensive examples from nature that the ratio of the golden section (.618...) and the Fibonacci series are associated with growth and self-organization. [By applying Seike's insight (reference 47) that activating a three dimensional projection of a natural, hyperspatial form actualizes that form, the following speculation arises: The projection of a vortex oriented orthogonally to our 3-space, appears as the logarithmic spiral whose characteristic rate of expansion, or contraction, is in the ratio of the golden mean. This may explain why Schauberger's (reference 42) logarithmic imploding vortices would interact and perhaps orthorotate the ZPE flux from an induced natural hyperspatial resonance. A waveform which rapidly changes frequency such that the length between adjacent zero crossings is in the ratio of the golden mean, called the "Fibonacci Chirp,", may be an activator of a natural, hyperspatial resonance.

54. M. B. King, "Stepping Down High Frequency Energy," Proceedings of the First International Symposium on

Non-Conventional Energy Technology, pp. 145-158, Toronto (1981).
This is an analysis of T. H. Moray's radiant energy invention (reference 52) from a systems theory point of view. In a non-linear system where energy is synchronously shifted down the spectrum, it is possible to integrate low amplitude, high frequency oscillations into large amplitude, low frequency oscillations. True synchronous electrical oscillations cannot be accomplished by moving electrons since their displacement is so large that numerous collisions would occur per oscillation cycle. However, the large mass of the ion limits its displacement allowing many collision-free cycles. Moray used ionic oscillations in his plasma tubes to step down high frequency energy.

55. T. E. Bearden, *Fer-De-Lance: A Briefing on Soviet Scalar Electromagnetic Weapons,* Tesla Book Co., Millbrae, CA (1986).
The author claims the Soviet Union has developed scalar electromagnetic weapons and offers evidence of their testing. Note 62 (pp. 107-108) stresses the importance of the rise time on the pulse creating the opposing electromagnetic fields and shows the orthorotated energy content is proportional to the square of the time derivative of the pulse.

56. L. L., P. H. Matthey, "The Swiss ML Converter—A Masterpiece of Craftsmanship and Electronic Engineering," in H. A. Nieper, *Revolution in Technology, Medicine and Society,* 'MIT' Verlag, Odenburg (1985).
The author describes a machine consisting of two oppositely charged, counter-rotating, acrylic disks separated by a smaller insulating disk and activated by pulsed magnetic fields. Over 3 KW at 230 Volts DC are extracted using brushes at the edge of the disks. Gravitational anomalies are also claimed. [In the air gap between the disks are two counter-rotating, electrostatically and oppositely charged, plasma vortices. Near the edge of the disks, the vortices are subject to abruptly pulsed, bucking magnetic fields which may orthorotate the hyperspatial ZPE flux.]

(See illustration on next page.)

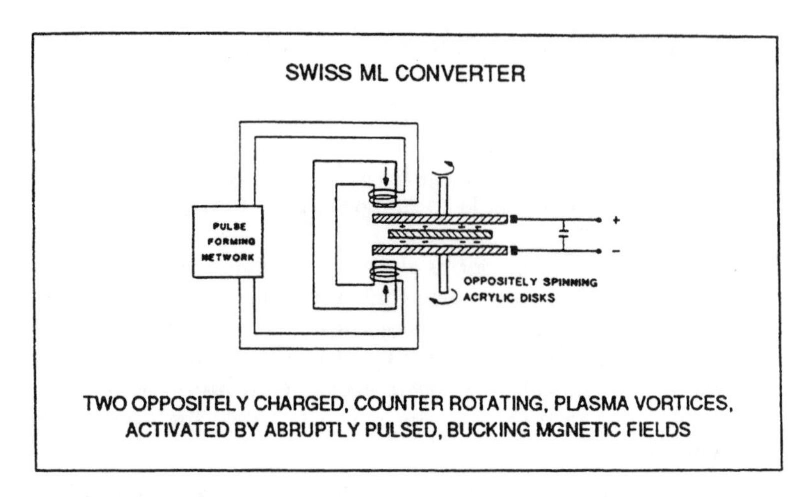

SWISS ML CONVERTER

TWO OPPOSITELY CHARGED, COUNTER ROTATING, PLASMA VORTICES, ACTIVATED BY ABRUPTLY PULSED, BUCKING MGNETIC FIELDS

57. J. Tennenbaum, "The Coming Breakthroughs in Biophysics," *Fusion* (Sept.-Oct. 1985), pp. 20-26.
The author discusses negentropic (self-organizing) processes including vortices, plasma vortex filaments, DNA radiation, structured water and the organization of the biosphere. He relates a negentropic model of DNA to the projection of a conic vortex onto a plane to manifest the golden mean, logrithmic spiral.

58. G. Zukav, *The Dancing Wu Li Masters,* Bantam Books, NY (1980).
This book is a layman's introduction to modern physics including the theory of relativity, quantum mechanics and an excellent exposition of the EPR paradox where distant events seem to be linked.

59. J. Gribbin, *In Search of Schrödinger's Cat,* Bantam Books, NY (1984).
This book is a layman's explanation of the development of quantum mechanics including an excellent description of Everett's many world's interpretation (reference 50).

60. J. P. Briggs, F. Peat, *Looking Glass Universe,* Simon & Shuster, NY (1984).
The authors describe for the layman the emerging theories of wholeness occurring in physics, chemistry, biology and neurophysiology where self-organization and nonlocal connectiveness appear.

61. R. Azevedo, P. Graneau, C. Millet, N. Graneau, "Powerful Water-Plasma Explosions," *Phys. Lett.* A 117 (2), 101-105 (1986).

This experiment measures the force imparted to a weight by an explosive electrical discharge in water. The magnitude of the force is anomalously large and has yet to be explained. In this experiment, no attempt has been made to optimize the thrust imparted to the weight by shaping the water vessel. [An experiment of this nature could imply a ZPE coherence only if the energy imparted to the mass exceeded the energy stored on the capacitor. If the explosive discharge could create an ionic vortex in the water (see next reference), an energy anomaly might be demostrated.]

62. L. Schroedter, "Vortex Launcher," Private Communication (August 1986).

The following shaped vessel will guide an explosive electrical discharge into a plasma vortex. The thick-walled vessel should

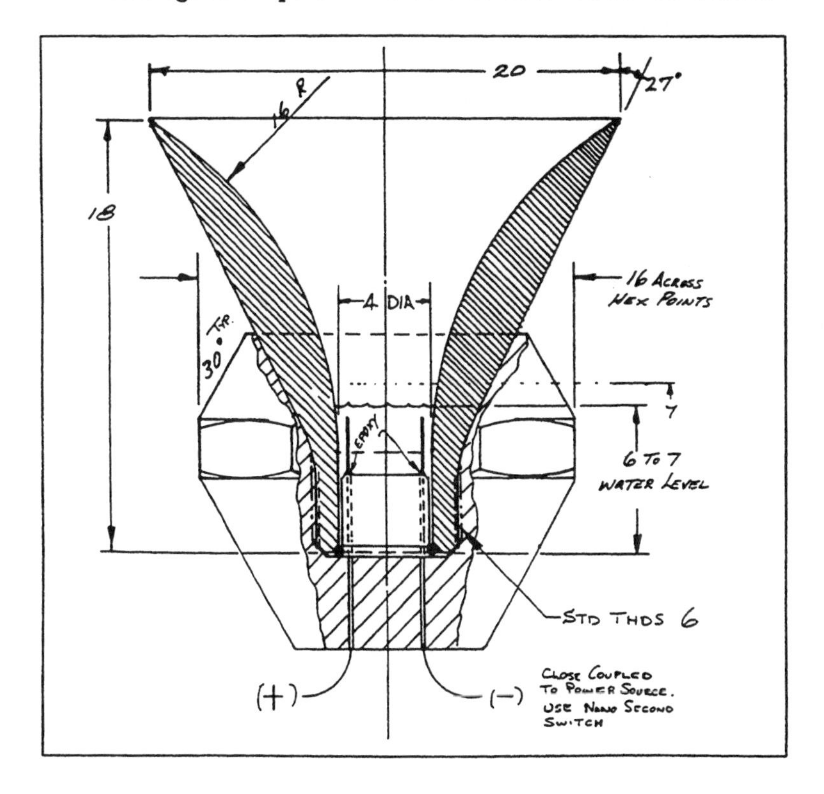

be machined from a strong ceramic (e.g. aluminum oxide). The electrodes are mounted off-center so that the explosive discharge induces a torque in the liquid.

63. D. Tichy, "Le Duc's Repeatable Production of Ball Lightning," Private Communication (August 1986).

    In the 1800's Stephane Le Duc repeatably produced with low power a self-contained electrical discharge that resembled ball lightning using a Wimshurst machine, Layden jar, and electrodes on a photographic plate or glass. See S. Le Duc, *Electric Ions and Their Use in Medicine,* Rebman Co., London, 1908, and *The Mechanism of Life,* Rebman Co., NY, 1914. [The repeatable production of ball lightning will allow the investigation of associated energy, time and gravitational anomalies. A gravitational anomaly can be demostrated by altering the period of a pendulum whose base consists of the apparatus producing the ball lightning. A gravitational anomaly implies a coherence in the ZPE since this is the only energy available to appreciably curve the space-time metric.]

THE HOLISTIC PARADIGM

July 1987

ABSTRACT

This review of holistic theories in science includes the many worlds interpretation of quantum mechanics, the zero-point energy, the EPR paradox, Bell's theorem, Bohm's implicate order, Prigogine's thermodynamics of self-organization, Sheldrake's morphogenetic fields, Pribram's holographic neurology and Woolf's holodynamic psychology. An integration of these theories yields implications for a new, expanded consciousness for humanity.

A new view of reality is emerging in Western science, a view that recognizes an inherent connectivity or oneness of seemingly separate objects; a view that allows consciousness, ultimately our very thoughts, to directly interact with other minds or objects. This view is called the holistic paradigm ("holistic" means whole; "paradigm" means pattern of belief). The concept of an inherent universal oneness is not new to humanity. The ideas have been expressed in ancient spiritual philosophies: Mysticism, Zen Buddhism, Taoism, Hinduism, etc. What is new is that Western scientists are gradually transforming their views, and simultaneously as they do so, a gradual transformation of humanity's consciousness is occurring. It will be shown that this consciousness transformation or "awakening" contains surprising implications for all of us.

REDUCTIONISM

To appreciate the emerging holistic paradigm of Western science, it is useful to understand the current reigning paradigm known as reductionism. The major belief of reductionism is that, in principle, any system can be understood by reducing it to parts and understanding their interactions. All interactions are always local (if certain parts are spatially separate, then a propagating signal mediates the interaction). The requirement that all interactions can be ultimately reduced to local interactions is known as the principle of local causality. Reductionism generally formulates its physical laws on the substrate of space-time. Modeling the universe with more than three dimensions of space is considered superfluous or unphysical. Reductionistic physics include classical physics (Newton's laws, Maxwell's electromagnetism, classical thermodynamics, etc.) and Einstein's theory of relativity. Nearly all of today's engineering and technology are based on reductionistic theories, and, consequently, most scientists are philosophically reductionistic.

QUANTUM MECHANICS

At the turn of the century, classical scientists declared that all the physical laws were known, and that only two "small clouds" were not adequately explained: the black body radiation spectrum of a heated object and the stability of electron orbits around atomic nuclei.[1] These two "small clouds" ushered in a scientific revolution—the development of quantum mechanics. Planck and Einstein quantized the radiation field giving birth to the photon, and Bohr, de Broglie and Schrödinger proposed quantized energy "standing waves" for the electron to explain atomic stability. The equations of quantum mechanics are not reductionistic in their implications. The probability waves described by these equations give rise to nonlocal interactions. This point was made by Einstein, Podalsky, and Rosen (EPR) in 1935 showing that for an atomic system where two, once-coupled particles separate, quantum mechanics predicts an instantaneous collapse of the wave function for either particle when its partner is measured—regardless of the distance between them.[2] In 1965, Bell reformulated Einstein's argument into a theorem that allowed the experimental investigation of this result.[3] Experiments were done in the 1970's confirming quantum mechanics,[4] but at the expense of undermining the principle of local causality.

Quantum mechanics begets other paradoxes as well. For example, in the two-slit experiment, an elementary particle exhibits wave or particle behavior depending on the detection apparatus. Wheeler has recently proposed a new configuration of this experiment where the detecting apparatus (for a wave or particle) is selected well after the particle/wave has passed the two slits, accenting the paradox.[5] Quantum mechanics shows that before a measurement is made, the particle/wave does not appear to have an objective, three-dimensional existence. But then what is it? The Copenhagen interpretation of quantum mechanics simply states that this is an irrelevant question—no attempt should be made to model the elementary particle/wave in a local way. Most physicists subscribe to this "interpretation" and give up attempts to explain what is going on underneath to give rise to the equations of quantum mechanics. This "interpretation" is perhaps the last bastion of reductionistic philosophy, which, in effect, simply ignores the astounding, hyperphysical implications of quantum mechanics.

MANY WORLDS

In the late 1950's, Everett introduced a totally self-consistent interpretation of quantum mechanics with a single assumption: That all systems—even the entire universe—are ultimately quantum mechanical in their nature.[6] This assumption had the advantage of resolving "the measurement problem." The measurement problem relates to the collapse of the wave function when a quantum mechanical event is recorded. The philosophical problem is, "where in a system does the quantum world end and the classical world begin?" By allowing the entire universe to be a quantum system, the equations of quantum mechanics then show that what really exists is an infinite number of three-dimensional universes coexisting simultaneously, and these can influence each other through waves of probabilistic coupling.[7] Our consciousness weaves a path through these many universes and experiences the perception of a single three-dimensional universe moving in time.[8] Perhaps what is most bizarre about Everett's many worlds interpretation is that within many of these universes simultaneously exists a multiplicity of ourselves. This notion seems absurd and perhaps is the reason that the many worlds interpretation is currently unpopular. However, as will be shown later, some novel concepts coming from the field of holodynamic psychology may make the notion of "self-multiplicity" not only less absurd, but perhaps even reasonable!

The many worlds interpretation introduces a hyperspace embedding an infinitude of three-dimensional spaces. The nonlocal EPR interaction is simply the selection by our consciousness of a particular three-dimensional universe. This idea can support the many works on positive thinking[9] where the process of visualization and emotional intensity (perhaps augmented by the alpha brain wave state)[10] allows the selection of which universe is more likely to be experienced from the set of probabilistic universes. The collapse of the quantum mechanical wave function is a selection of a universe by our consciousness. This could be the most powerful and optimistic implication coming from quantum mechanics, for it gives us more choice than we have ever dreamed possible.

ZERO-POINT ENERGY

Quantum mechanics has also discovered the existence of an all-pervading energy embedded within the fabric of space, the zero-point energy. About thirty years after the

Michelson and Morley experiment failed to detect the ether, it was recognized that a term was needed in the equations of quantum mechanics in order for them to correctly describe experimental results. The term described an inherent, electrically energetic fluctuation interacting with all systems, even in the total absence of mass, radiation or heat. Zero-point refers to a temperature of absolute zero degrees Kelvin and means that the fluctuations are not thermal in nature. The zero-point fluctuations at first presented an embarrassing situation—their energy density was infinite.[11] Mathematical procedures called renormalization were devised for quantum mechanical calculations to remove this infinite zero-point energy to yield the finite masses and fields that we observe. A new and even more promising approach to modeling the zero-point energy may arise from superstring theories which unify all the forces of nature.[12] Recent experiments have shown that the zero-point energy is unaffected by the presence of absorbers and reflectors.[13] This implies that the energy does not propagate in our three-dimensional space. Where does it come from?

By applying the formalism of general relativity to the zero-point energy, Wheeler's geometrodynamics answers this question and derives the modern view of the fabric of space.[14] In general relativity, a sufficiently large energy density causes the fabric of space-time to pinch (like a black hole) in a direction orthogonal to our three-dimensional space. This yields hyperspatial channels Wheeler calls *wormholes* through which this energy flows. Wormholes can link distant locations within our universe or create linkages to other parallel, three-dimensional universes. (This hyperspatial description has similarities to Everett's many worlds interpretation.)[15] The action of the zero-point fluctuations results from an electric flux orthogonally passing through our three-dimensional space. Jitter in this flux, aligned with our three-space, gives rise to a turbulence of microscopic white holes (electric flux entering) and black holes (flux leaving) that are constantly pair-forming and pair-annihilating. These holes are on the order of 10^{-33}cm and the resulting turbulence gives the fabric of space a dynamic foam-like structure sometimes called the *quantum foam.*

The electrical flux through any of these channels has a (mass-equivalent) energy density on the order of 10^{93}grams/cm^3. This is simply enormous compared with the masses of elementary particles or even the Earth (10^{12} grams). Bohm suggests that the zero-point energy is the implicate source of all explicate matter and energy. But, how can a sea of turbulence give rise to the coherent structures that we observe as matter?

SELF-ORGANIZATION

The reductionistic view is that a random chaotic system should remain so. This is the common understanding of the law of entropy, the second law of thermodynamics. This view applies to those systems which are linear or near equilibrium. However, there are other types of systems.

In 1977, Ilya Prigogine won the Nobel prize in chemistry for identifying under what conditions a system may evolve from a chaotic state to an organized state. The conditions are that the system must be nonlinear, far from equilibrium, and have an energy flux through it.[16] A nonlinear system is one whose response to a set of stimuli may produce new, surprising or synergistic behavior that cannot readily be predicted by simply summing the responses of the individual stimuli. (Linear systems, on the other hand, exhibit linear superposition where a reductionistic "sum of parts" view gives the correct prediction. In the history of science, most systems that were readily analyzable were linear systems. For these systems, the reductionistic view works quite well.) Examples of self-organization is the generation of vortices and vortex rings in a turbulent fluid or air (e.g., tornados). Another example of order arising from chaos occurs in a turbulent plasma (highly ionized gas). Here, the formation of vortex ring pairs is observed to occur.[17] This is of interest since the zero-point energy can be modeled as a virtual turbulent plasma. Vortex ring formation would then correspond to pair production of elementary particles (e.g., electron and positron). Note that a vortex ring exhibits a precessional rotation (a helical rotation around a cylinder that closes into a toroid). Many investigators have suggested that a precessional rotation may be a key for orthorotating into our three-space the hyperspatial flux that constitutes the zero-point energy.[18] From a system's perspective, the behavior of the zero-point energy fulfills the conditions for self-organization. It is highly nonlinear in its interaction with matter; it can be driven far from equilibrium by abrupt motions of matter (or plasma); and it is maintained by a (hyperspatial) flux of electrical energy.

The holistic paradigm has the zero-point energy as the source that maintains the elementary particles and, therefore, all matter. It has recently been shown to be the basis for the stability of the hydrogen atom.[19] Bohm shows that it is the basis of the implicate order from which arises the explicate phenomena of matter, energy, time and space.[20] Bohm's implicate order contains a quantum potential that results in nonlocal correlations across space-time[21] (as well as perhaps across the multiple, parallel universes of Everett). These nonlocal linkages result in a holistic description of our universe as a hologram—where the whole view is implicitly embedded in every section of the hologram. Note that Wheeler's hyperspatial wormholes likewise generate a nonlocal connectivity. The zero-point energy constitutes the first substrate of organization, and allows phenomena to be linked nonlocally through a higher dimensional space.

MORPHOGENETIC FIELDS

Sheldrake has proposed the existence of subtle, hyperspatial "morphogenetic fields" which guide the formation of matter or living systems.[22] These fields are further strengthened by the physical manifestation they help form, thus making it easier to repeat creating the physical form. For example, in chemistry it is often very difficult to grow a new crystalline compound for the first time; but, after one laboratory succeeds, it is easier for others to accomplish this, even at remote locations. A process that was previously failing begins to succeed after the first success. Similarly, the creation of predicted, new, elementary particles in an accelerator is difficult at first, but once the new particle is created, it is recreated easier in accelerators all over the world (even under the old experimental conditions). In Sheldrake's theory, the crystal or elementary particle has a morphogenetic field which becomes "locked in" at the first physical manifestation. This field then guides future growth and creation. The field is nonlocal and hyperspatial in its nature and can be likened to an "etheric" or "spiritual" form.

Sheldrake's theory applies especially to biological systems, and here the morphogenetic fields can give rise to group mind or collective intelligence. In embryology, the embryo of a dragon fly that is cut in half still yields a fully formed dragon fly—except that it is half size.[23] The fields guide the morphogenesis of the embryo, as opposed to only an internal, localized, reductionistic, growth mechanism. Thomas shows an example of collective behavior occurring in a single cilia of a protozoan.[24] The cilia itself is a colony of separate microscopic organisms that combine to produce a single unitary filament. Hundreds of cilia are synchronized to propel the protozoan. Another example is the mitochondria. They live within the protoplasm of a single cell as autonomous beings—yet they participate in a collective fashion to provide the cell with energy.[25] The morphogenetic fields of a single cell guide the collective behavior of its components.

The morphogenetic fields may also link the separate individuals of a specie. An example of collective intelligence occurs in the insect world with termites. When there are only a few termites, their pattern of building or moving pellets is random and meaningless. Yet, as more termites are added to the group, a threshold phenomena occurs where their behavior radically changes and they begin to cooperatively create majestic, multi-arch structures for their nest.[26] Another example occurs in squid migration. With one or just a few squid gathered, there is no awareness of what direction to swim; but when a sufficient number are present, a new group intelligence arises, and the collective acts as a single organism making a direct, purposeful migration across the ocean.[27] Like the clear image that can be achieved through a large holograph versus the noisy image through a small one (or piece), a large collective of individuals is needed to manifest a clear intelligence.

Pribram makes this point, as well, in his holographic theory of memory storage in the brain.[28] Here, memory is stored redundantly on many neurons. The ability, clarity and quickness of recall is related to the large number of neurons. Experiments have shown that memory is not localized in the brain, but redundantly distributed.

Transgenerational collective intelligence has been shown in experiments with mice. Since the 19th Century, a particular specie of mice has been used in psychological experiments where the mice have been taught to run mazes. It has been observed that the later generations are able to learn faster to run the mazes.[29] Here, intelligence is associated with the morphogenetic field of the species, and each individual is able to resonate with this field benefiting from, and adding to, the group intelligence.

Perhaps the most famous example of species collective intelligence was observed in monkeys on the Pacific islands near Japan.[30] While studying the behavior of the monkeys, the scientists noted that they refused to eat sweet potatoes because of the sand on them. A scientist taught one of the monkeys to wash a sweet potato and it began to consistently wash and eat them. Soon, by imitation, other monkeys on the island began washing their sweet potatoes. In a few weeks, all the monkeys on the island had learned to wash them. Now, the big surprise came when the scientist sailed to another island inhabited by the same species of monkey. When they arrived, they observed that all the monkeys on this separate island were washing their sweet potatoes as well! It was as if this knowledge became encoded in the collective group mind of the species, and this mind (or morphogenetic field) was nonlocal within space-time, yet each member of the species is part of it.

HOLODYNAMIC PSYCHOLOGY

Holistic psychology extends the concept of a nonlocal collective group mind to human beings. Woolf describes the process as "holodynamic" psychology since all minds are in a constant dynamic state of growth, yet all are a part of the group collective or holistic universal mind.[31] The universal mind exhibits the following recursive, archetypical process: it gives rise to many individual human minds, each experiencing a separation from the universal mind. Each human mind, in turn, is comprised of many more primitive minds called "holodigms"—each with its own ego that experiences separation from the other holodigms. What we experience as our ego is simply the holodigm that is currently active or conscious. The word "holodigm" means whole (holo), form (digm). It implies that each primitive ego state is a form that arises from the holistic, universal mind and contains the potential for reconnecting its awareness back to the universal mind. The process for establishing this reconnection or awakening, is called psychomaturation. This process not only yields a happier, more fulfilled life, but also unlocks the psychic potential of the individual.

Psychic, extrasensory abilities such as telepathy, psychokinesis, astral traveling, precognition, cognition of other former lives, etc., arise simply by expanding one's identity and awareness into the universal mind.[32] In the past, it would normally take many years of mystic training, meditation, and practice to begin to achieve this awareness. The process of psychomaturation accelerates this awakening by removing the blocks that inhibit this and, perhaps most importantly, by achieving a bonding experience with others on this same path of growth. When many minds are focused together in the psychomaturation processes, the awareness accelerates not only for those experienced, but for those who are just beginning the process. The more bonded minds participating, the more rapid the growth. Thus, as more people awaken to their full potential selves (the spiritual-self which transcends the physical body), the easier it will be for others to awake. When a sufficient number come into experiential awareness of the universal mind, a threshold will be breached in the morphogenetic field of mankind, and all minds will spontaneously become universally aware. At this point, all individuals will realize and directly experience that we are a single superconscious entity.

SUMMARY

Quantum mechanics has given a new view of reality to Western science. Perhaps the biggest surprise to the reductionistic view is the existence of nonlocal connectivity. The successful experimental demonstration of the EPR paradox is the "crack in the cosmic egg"[33] from which is emerging the new holistic paradigm. Quantum mechanics also shows that every elementary particle and, therefore, all matter is formed in the zero-point energy which exhibits a nonlocal or hyperspatial quality. Bohm proposes an implicate order in the zero-point energy, and Sheldrake suggests the existence of subtle, hyperspatial morphogenetic fields which guide the hierarchical organization of matter and living systems. Thomas observes this group organization and group intelligence throughout biology, and Woolf has developed a process to accelerate the experiential awareness of our universal mind. It is hoped that this brief overview motivates the study of the cited references, for a growing awareness of the holistic paradigm will usher in a unifying transition for humanity.

REFERENCES

1. G. Ganow, *Thirty Years That Shook Physics,* Doubleday, NY, 1966. This layman's text gives the historical development of quantum mechanics. It contains a picturesque description of Dirac's virtual electron-positron sea as a model for the zero-point vacuum energy.

2. G. Zukav, *The Dancing Wu Li Masters,* Bantam Books, NY, 1979. This layman's overview of modern physics contains a thorough discussion of the EPR paradox.

3. J. F. Clauser, A. Shinony, "Bell's Theorem: Experimental Tests and Implications," *Reports Prog. Phys.* 41, 1881 (1978). Bell's Theorem presents the EPR paradox in a quantitative, falsifiable manner which allows experimental investigations. The experiments have confirmed the paradox.

4. Ibid.

5. J. A. Wheeler, "Beyond the Black Hole," in H. Woolf (ed.), *Some Strangeness in the Proportion,* Addison-Wesley, Reading, Mass., 1980; pp. 341-375. Wheeler discusses the implications of quantum mechanics and shows how the observer is inseparable from the observation.

6. H. Everett, "Relative State Formulation of Quantum Mechanics," *Rev. Mod. Phys.* 29(3), 454 (1957). Also, B. S. DeWitt, N. Graham, *The Many Worlds Interpretation of Quantum Mechanics,* Princeton University Press, 1973. Everett introduces a self-consistent formulation of quantum mechanics without the necessity of invoking postulates regarding the observer. The formulation yields a superspace containing an infinite number of three-dimensional spaces.

7. P. Davies, *Other Worlds,* Simon and Schuster, NY, 1980. This book describes the various views of superspace arising in modern physics, including the many worlds interpretation of quantum mechanics.

8. J. Gribbon, *In Search of Schrödinger's Cat,* Bantam Books, NY, 1984.

   Gribbon presents the historic developments of modern physics for the layman, including a thorough explanation of the many worlds interpretation of quantum mechanics.

9. C. M. Bristol, *The Magic of Believing,* Pocket Books, NY, 1969. This book teaches how visualization of desirable outcomes helps to trigger those events.

10. J. Stearn, *The Power of Alpha-Thinking,* Signet, NY, 1977. Stearn shows that visualization of desirable events while in an alpha brain-wave state increases their likelihood.

11. E. G. Harris, "The Problems of Infinites in Quantum Electrodynamics," in *A Pedestrian Approach to Quantum Field Theory,* Wiley Interscience, Chap. 10., NY, 1972.

In chapter 10 is a discussion of the zero-point energy, including experiments that have detected its existence.

12. M. Kaku, J. Hainer, Beyond Einstein: *The Cosmic Quest for the Theory of the Universe.* Bantam Books, NY, 1987.
This book is a layman's introduction to the various unification theories of modern physics. Superstring theory appears to be the most promising for unifying all four fundamental forces (strong nuclear, weak nuclear, electromagnetic, and gravitation). The fundamental unit in this theory is a filament on the order of 10^{-33}cm (same size as Wheeler's quantum wormholes) that can form into a Mobius loop. The theory avoids the problems of renormalization by not using point particles.

13. O. H. Abroskina, G. Kh. Kitaeva, A. N. Penin, "The Effective Brightness of Zero-Point Fluctuations of the Electromagnetic Vacuum of Parametric Scattering of Light," *Sov. Phys. Dokl.* 30(1), 67 (1985).
This paper summarizes an experiment that shows that zero-point fluctuations are independent of the presence of nearby reflectors or absorbers. This shows that the zero-point energy does not arise from a propagation in our three-dimensional space, and thus supports the hyperspatial flux model for the source of the zero-point energy.

14. J. A. Wheeler, *Geometrodynamics.* Academic Press, NY, 1962.
Wheeler derives the modern view of the fabric of space by applying general relativity to the zero-point energy of quantum mechanics. The result is a dynamic maelstrom of nonlocal, microscopic channels across space-times through a turbulence of "wormholes" whose size is on the order of 10^{-33}cm. An electric flux passes through these channels with an energy density on the order of 10^{93} grams/cm^3.

15. B. Toben, F. Wolf, *Space-Time and Beyond,* E. P. Dutton, 1982.
This book is an illustrated introduction to the theories of superspace, geometrodynamics, the many worlds interpretation and the resulting possibilities for expanding consciousness.

16. I. Prigogine, I. Stengers, *Order Out of Chaos,* Bantam Books, NY, 1984.

Prigogine overviews his Nobel prize-winning theory for the layman, showing the conditions under which a system may evolve from a chaotic state to an organized one.

17. W. H. Bostick, "Experimental Study of Plasmoids," *Phys. Rev.* 106(3), 404 (1957).
Bostick experimentally investigates the formation of plasmoid vortex ring pairs in a turbulent plasma. A "quantum condition" in the ratio of toroidal to poloidal diameters is identified as needed for stability.

18. M. B. King "Cohering the Zero-Point Energy," Proceedings of the International Tesla Symposium, International Tesla Society, Colorado Springs, CO, 1986.
By merging theories of the zero-point energy with theories of system self-organization, it is shown that it may be possible to cohere the zero-point energy as a source. Possible methods are referenced and include ion-acoustic oscillations, abruptly pulsed bucking fields, and plasma vortices and vortex rings exhibiting precession.

19. H. E. Puthoff, "Ground State of Hydrogen as a Zero-Point Fluctuation Determined State." *Phys. Rev* D 35(10), 3266 (1987).
The author shows that the ground state of the hydrogen atom results from an equilibrium between radiation emitted due to acceleration of the electron and radiation absorbed from the zero-point fluctuations.

20. K. Wilbur, *The Holographic Paradigm*, Shambhala, Boulder, CO, 1982.
Wilbur combines a series of articles and interviews with the proponents of the holographic paradigm in Western science: David Bohm in physics and Karl Pribram in neurology. Bohm has proposed an implicate/explicate order with the zero-point energy at its basis. This gives rise to a holographic universe. Pribram presents a holographic model for the brain. Combining their work gives a holographic processing brain interacting with a holographic universe which yields the explicate perception of a space-time universe.

21. D. J. Bohm, B. J. Hiley, "On the Intuitive Understanding of Nonlocality as Implied by "Quantum Theory," *Found. Phys.* 5(1), 93 (1975).
The authors show how nonlocality arises in quantum mechanics through the EPR paradox and Bell's Theorem. Bohm's "quantum potential" leads to the notion of an unbroken wholeness for the entire universe.

22. T. P. Briggs, F. Peat, *Looking Glass Universe*, Simon & Shuster, NY, 1984.
This book is an excellent overview of the holistic theories emerging in Western science. It includes Bohm's implicate/explicate order, Prigogine's thermodynamics, Sheldrake's morphogenetic fields, and Pribram's holographic neurology. These holographic theories suggest a nonlocal connectiveness linking all of nature.

23. Ibid.

24. L. Thomas, *The Lives of a Cell*, Penguin Books, NY, 1978.
This book contains a collection of essays by Thomas that show the remarkable hierarchical organization at various levels in biology. The essay, "Some Biomythology," shows the cilia of a certain protozoan are a colony of spirochetes.

25. Ibid.
The essay, "Organelles as Organisms" show that the mitochondria are autonomous organisms within the cell.

26. Ibid.
The essay "Living Language" describes the observation of termites' ability to build majestic structures when in large numbers.

27. Ibid.

28. Briggs and Peat, *Looking Glass Universe*, Op. Cit.

29. Ibid.

30. Ibid.

31. V. Woolf, *Holodynamics*, Harbinger House, NY, 1990.
The author presents a mind model based on levels of maturation which shows that in our fullest potential our minds link to a universal mind. Specific techniques for psychomaturation are presented with numerous case histories. The maturation process accelerates when many minds are focused in a bonding experience. Woolf extends the holistic concepts of Bohm and Pribram into a "holodynamic" psychology.

32. D. Loye, *The Sphinx and the Rainbow*, Shambala, Boulder, CO, 1983.
This is a thorough overview of consciousness, brain and mind research. It shows that the brain's frontal lobes are associated with bonding, psychic abilities, mystical awareness, and linking to a universal mind.

33. J. C. Pearce, *Exploring the Crack in the Cosmic Egg,* Pocket Books, NY, 1975.
This scientific and philosophical study examines the nature of paradigm shifts through human history. The author supports with numerous references the notion that our objective reality is created by a universal mind in which we all participate. This can allow paranormal phenomena.

DEMONSTRATING A ZERO-POINT ENERGY COHERENCE

July 1988

ABSTRACT

The notion of a physical hyperspace frequently arises in modern physics. The zero-point energy can be modeled as an electric flux from the fourth dimension intersecting our three dimensional space. It manifests as a turbulent, virtual plasma. The observation of self-organizational modes in plasmas suggests experiments that may cohere the zero-point energy and produce corresponding gravitational anomalies. The suggested experiment uses sharply pulsed, bucking magnetic fields produced within a caduceus coil whose core is a plasma tube resonating in the ion-acoustic mode.

INTRODUCTION

A paradigm shift is occurring right now in the field of physics. Quantum mechanics has ushered in a new, holistic view of our universe which can allow nonlocal linkage of distant events in seeming violation of the common sense principle of local causality.[1] This result has been confirmed by experiments.[2] In addition, quantum mechanics shows the existence of an all-pervading energy existing in the fabric of space called the zero-point energy (ZPE).[3-5] Recent advances in thermodynamics and theories of system self-organization open the possibility for tapping this energy as a source.[6-10] This paper explores how to demonstrate this experimentally.

Most scientists today believe the ZPE cannot be tapped as a source. In fact, there is a paradox concerning how much of this energy even exists.[11] The successful standard model for elementary particles as well as quantum electrodynamics requires virtually an infinite amount of ZPE imbedded within each point in the fabric of space. Here there occur rapid, tremendous fluctuations of electric flux which interact with every elementary particle. Yet the net manifestation of how much energy that is "really" there seems quite small and is difficult to detect. How can an infinity be imbedded in a point? A corresponding question is "from where do the zero-point fluctuations arise?" Recent experiments show that the ZPE action in a region is independent of nearby reflectors and absorbers showing that the energy is not a radiation field like light propagating through space.[12] This question is similar to asking from where does an elementary charge's electric flux arise? The answer takes us to the heart of the current paradigm shift in physics, for it implies the existence of what most scientists and laymen generally do not believe: There exist more physically spatial dimensions to our universe than the three dimensions of space (length, height, and width) that we perceive. Could it be that the three dimensional Euclidean world moving in time, which we intuitively assume to exist, is actually an artifact of our consciousness? Are we like "flat landers" who in their two dimensional universe had no experience or conception of how a third spatial dimension could possibly exist? Mystical thought and the ancient Eastern religions (e.g. Buddhism, Taoism, Hinduism, etc.) have always professed that our everyday three dimensional world is an illusion or a subset of a greater reality whose description is beyond words. Modern quantum mechanics delivers the same message with interpretations ranging from "it can not be physically modeled" (Copenhagen interpretation), to "there exists an infinite number of simultaneous three dimensional universes" (Everett's many worlds interpretation[13-15]), to "all explicate phenomena arise from a hidden, nonlocal, implicate order" (Bohm's quantum potential[16]). The experiments of quantum mechanics have conclusively shown that there is more going on than can be modeled from a three dimensional perspective.

HYPERSPACE

Postulating the existence of just one more physical dimension creates resolutions to some of the philosophical questions arising in physics. For example, why does a charged particle's electric field decay at exactly an inverse distance squared ($1/r^2$) ratio (i.e., why is the exponent exactly equal to two?). Nineteenth century scientists modeled the electric field as a uniform fluid flow with the charge as the source. Here the electric flux would distribute itself uniformly on the surface of an imaginary sphere surrounding the charge. Since the sphere's surface area is proportional to the square of the radius, the $1/r^2$ exponent is exactly two; but from where does this electric flux arise? The source of the flux can be modeled as arising from a physical fourth dimension. In Figure 1, our three dimensional space is represented by a

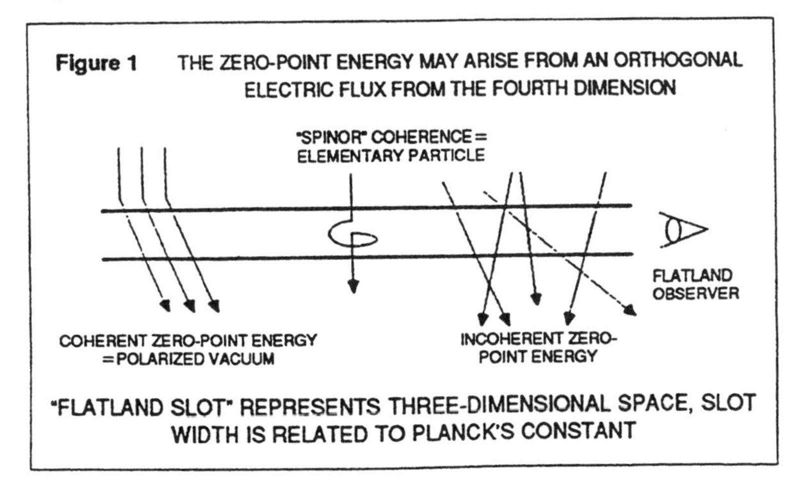

Figure 1 THE ZERO-POINT ENERGY MAY ARISE FROM AN ORTHOGONAL ELECTRIC FLUX FROM THE FOURTH DIMENSION

"flatland slot" which has a thickness in the fourth dimension proportional to Planck's constant.[17] Electric flux flows orthogonally through our 3-space. The manifestation of this flow in our 3-space begets a turbulence of mini virtual charges whose scale is on the order of Planck's length, 10^{-33}cm.[18,19] This turbulence is sometimes called the "quantum foam." An elementary charge twists or orthorotates this flux into our three dimensional space. Later it will be shown how a vortex ring model for the elementary charge accomplishes this.

The rate of this orthogonal flux flow through our flatland slot is intimately related to the speed of light and the pace of time. From the theory of relativity all measurements of light's velocity in a vacuum beget a constant value while the pace of time is flexible. An experiment that altered the ZPE flux could influence the pace of time near the apparatus. The pace of time is just one component of general relativity's space-time metric. Since the metric is proportional to the stress-energy tensor, a change in the tensor via a ZPE coherence could give rise to curving the space-time metric yielding artificial gravitational fields. Thus an experiment that slows the pace of time (e.g. by altering the frequency of a mechanical oscillator) or alters the weight of the apparatus would demonstrate the successful coherence of the ZPE.

This hyperspatial model of the ZPE gives a geometric interpretation to special relativity's description of the electric field. If an observer moved uniformly away from a stationary electric charge, he would then detect the existence of a magnetic field. Special relativity shows that the magnetic field is a Lorentz transformation of the electric field. In this case it can be made to appear or disappear depending on the motion of the observer. The relativistic transformation of force, fields or mass can be illustrated geometrically by a tilt in a Minkowski diagram (Figure 2). If

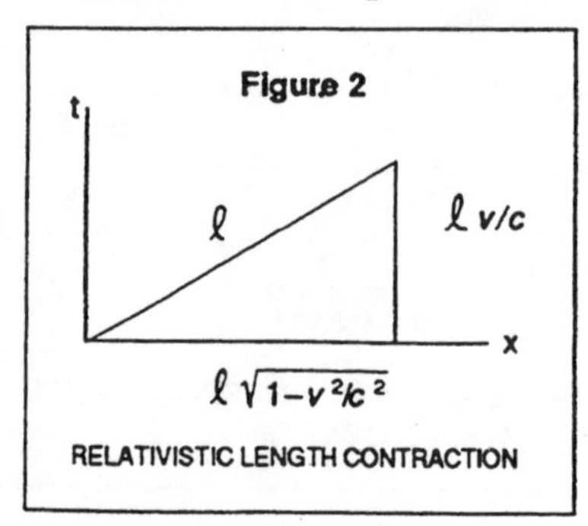

RELATIVISTIC LENGTH CONTRACTION

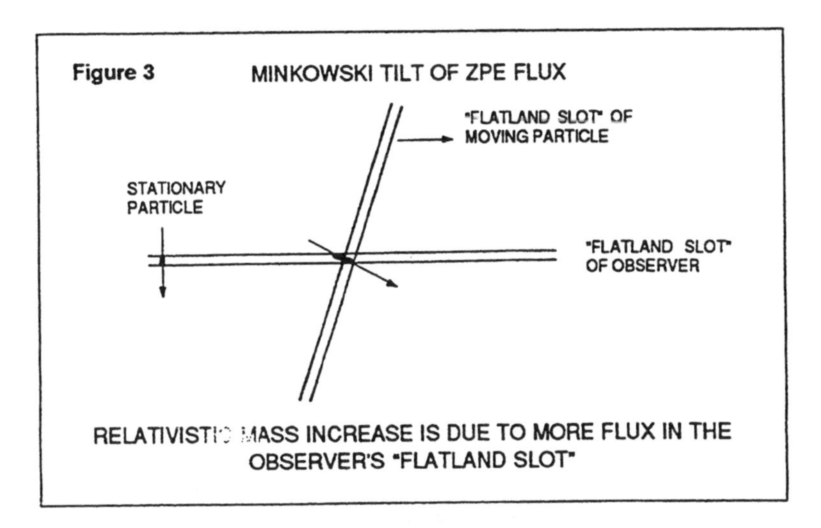

Figure 3 MINKOWSKI TILT OF ZPE FLUX

RELATIVISTIC MASS INCREASE IS DUE TO MORE FLUX IN THE OBSERVER'S "FLATLAND SLOT"

the Minkowski tilt angle applied to tilting the normal hyperspatial ZPE flux that "feeds" the moving field or mass, a component of this tilted ZPE flux would align in 3-space (Figure 3). Thus the moving observer sees a slightly orthorotated perspective of the electric field. This manifests as the magnetic field. The ZPE flux tilt can likewise be used to model the relativistic mass increase. Here the ZPE flux maintains the mass of an elementary particle like a flowing stream maintains a vortex. The moving observer sees more of this flux aligned in his 3-space, and it manifests as a mass increase on the particle (Figure 3). The hyperspatial ZPE flux model gives a physical explanation of relativistic transformations.

Another problem in relativistic physics involves modeling light. An intriguing property of light is that it apparently cannot be successfully modeled as a propagating three dimensional wavefront. Special relativity shows this in a textbook derivation of the Lorentz transformation.[20] To illustrate, imagine two observers, one stationary and the other moving at constant velocity close to the speed of light (Figure 4). At the instant their positions align, a flashbulb ignites and launches a spherical wavefront starting from the origins of both their reference frames. After a period of time, their origins are separated. If each observer had apparatus set up

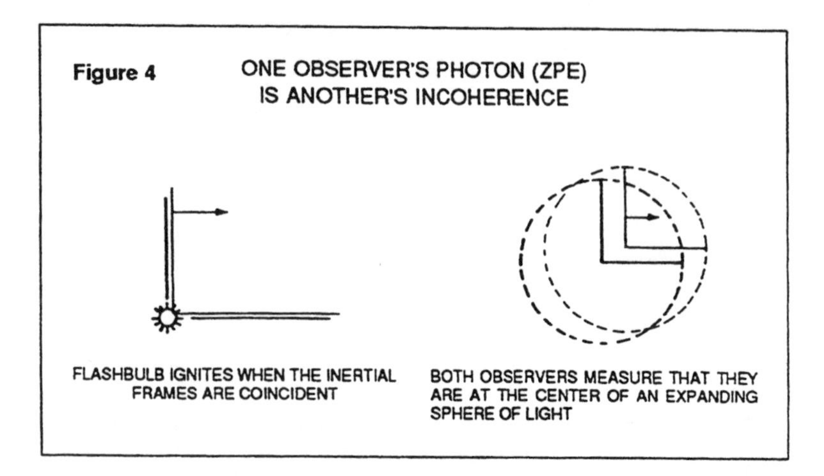

Figure 4 ONE OBSERVER'S PHOTON (ZPE) IS ANOTHER'S INCOHERENCE

to detect this expanding wavefront at a later time, each would see himself at the center of an expanding spherical wavefront—yet their centers are separate. Light cannot be a single expanding spherical wavefront. In this case, how can light be multiple expanding wavefronts for all observers?

A hyperspatial model can resolve this paradox. If light were a propagating orthorotation of the fourth dimensional ZPE flux in each observer's universe with the angle of this flux tilted for the moving observer relative to the stationary observer (the tilt angle matches the tilt in a Minkowski diagram), then the action of the ZPE flux would yield separate expanding coherent wavefronts for each observer. In this model, the other observer's wavefront would manifest as incoherent, background zero-point fluctuations. A corresponding model for a photon would be an expanding toroidal form that remains connected in the higher dimensional space despite its nonlocal expanding character in 3-space.[21] When the photon is absorbed in an atomic system, the coherence of this toroid is immediately disrupted returning it to an incoherent ZPE manifestation. This photon model contains both particle and wave manifestations and is similar to Bohm's description of a nonlocal, quantum potential arising from a hidden implicate order in the ZPE.

Similarly, a nonlocal connectivity can be modeled to explain the quantum connectiveness of the EPR paradox where two separate elementary particles ejected from a single atom remain correlated in their statistical behavior.[2] If each particle is maintained by an orthogonal ZPE flux, and the flux for each arises from a hyperspatial split flux stream that remains connected in the fourth dimension (Figure 5), then the particles remain correlated since they are a single hyperspatial object whose 3-space projection appears as separate particles. Figure 5 is similar to a Feynman diagram with the flatland slot representing an instant of time. In this model, the fourth dimensional object exists across time, while our consciousness or perception slices it into the projection of 3-space objects moving in time.

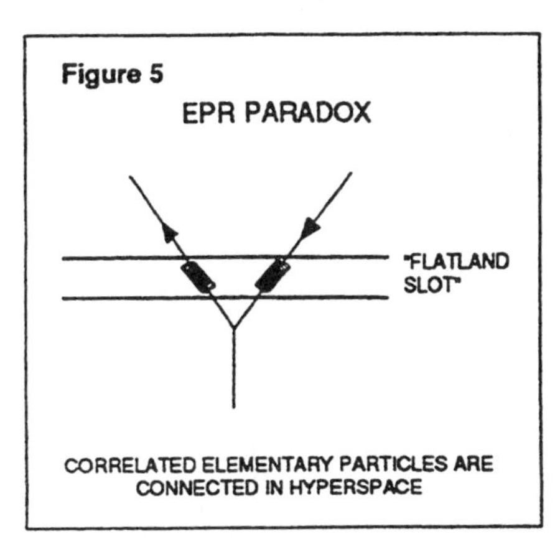

Figure 5
EPR PARADOX
CORRELATED ELEMENTARY PARTICLES ARE CONNECTED IN HYPERSPACE

The ultimate hyperspatial model of reality that is totally consistent with the equations of quantum mechanics is Everett's many worlds interpretation. Here an infinite number of parallel three dimensional universes exist across the higher dimension. What we perceive as our three dimensional universe is simply a continual 3-D projection perceived by our consciousness of the hyperspatial reality. This interpretation is appealing to the mystic for it allows choice for that universe and consequential experience which is most desired. This model supports the many works on positive thinking, affirmations, visualization, etc., in which conscious volition supported by strong emotions plays a role in selecting which probabilistic 3-D universe will be experienced. In support of this notion, there exists new and convincing experimental evidence that our minds influence probabilistic events.[22] Overall there is ample evidence in physics to support the existence of a physical hyperspace.

COHERING THE ZERO-POINT ENERGY

A hyperspatial flux model for the zero-point energy opens the possibility of tapping it as a source. In principle it becomes theoretically possible to cohere the ZPE when two separate areas of theoretical physics are combined. These are: Theories which model the ZPE as a virtual plasma, and theories of system self-organization. Prigogine won the 1977 Nobel Prize in chemistry for showing in thermodynamics under what conditions a system evolves from chaotic turbulence to a state of self-organization.[23] The conditions apply for any type of system: it must be nonlinear, far from equilibrium, and have an energy flux through it. The ZPE fulfills these conditions. It is highly nonlinear in its interaction with matter, it can be driven from equilibrium with abrupt motion or abrupt electrical discharges, and it arises from a hyperspatial electrical flux. Nonlinear hydrodynamic modeling of the ZPE has already shown certain modes yield a net energy gain.[24] Magnetohydrodynamic modeling could be even more fruitful, for by modeling the ZPE as a virtual plasma we can learn about its self-organizational properties by studying the behavior of plasmas.[25,26.]

A common self-organizational observation of plasmas regards the formation of helical filaments. (It is ironic that these were called "instabilities" in the early attempts to force quasi-linear plasma behavior in fusion experiments.) These filaments have a tendency to squeeze into an elongated string-like form which may in turn start another, higher order helical filament. The ZPE virtual plasma may also exhibit this property, recursively forming higher order helical filaments. This could provide a physical basis for superstring theories.[45] In a plasma, if a filament closes onto itself it produces a stable toroidal vortex ring called a plasmoid.[27] This form has been used to model ball lightning.[28] Similarly in the virtual ZPE plasma, the higher order filaments or strings that close into loops yield the elementary particles. The self-organizational behavior of plasmas has been well observed and, on a large scale, has recently been proposed as the dominant mechanism for the formation of galaxies.[29]

Plasmas may also provide the technological link for tapping the ZPE. A plasma's electrons and ion nuclei interact directly with the ZPE via vacuum polarization. Quantum electrodynamics shows that the vacuum polarization interaction is very different for the different elementary particles. [30] Electrons, especially electrons in a conductor, behave like a smeared charge cloud in thermodynamic equilibrium with the zero-point vacuum fluctuations. Nuclei, on the other hand, exhibit a stable vacuum polarization with lines of flux converging steeply onto these particles (Figure 6). Furthermore, the high mass-energy density of electric flux converging onto the nucleus can yield a stable, space-time metric curvature directly into the fourth dimension from which the ZPE flux flows. Thus the nuclei of a plasma's ions become an important component for orthorotating the ZPE flux.

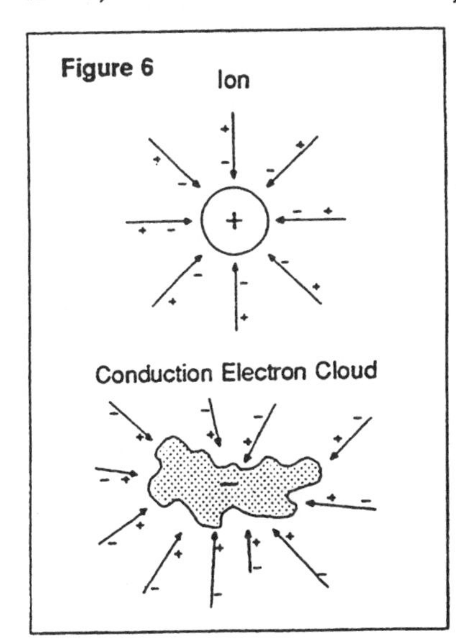

A ZPE-plasma synergy may arise during the collective behavior of the ions in a plasma. Ion oscillation in a plasma is known as the ion-acoustic mode, and it has been experimentally observed to give rise to energetically anomalous behavior (e.g. run-away electrons, high frequency spiking, anomalous heating, etc.[31]). T. H. Moray stressed the importance of ion oscillations in the plasma tubes of his well-documented invention that produced anomalously high power.[32]

The creation of ion helical filaments, vortices and vortex rings may also yield a ZPE orthorotation. The flying disks of Searl and Carr, the Swiss ML converter, and Gray's motor exhibit helical, radial plasma discharges along their segmented rotors.[33] When rapidly spun, these helical discharges bend into curved spiral filaments which collectively form an ion vortex (Figure 7). Each bent spiraling filament is a section of a plasmoid. The plasmoid vortex ring exhibits precessional motion of the plasma particles—a poloidal rotation around the filament closing into a toroidal rotation yielding two orthogonal spins. Forced precession has been suggested as a method for orthorotating the ZPE flux into our 3-space.[34] It may be the mechanism that allows ball lightning to persist. One could also pump a charged fluid, such as mercury, through a helical toroid piping system to create a flowing ion vortex ring exhibiting ionic precession. Seike[35] and Cho[36] have also proposed electrical precessional schemes where the 3-space projection of a fourth dimensional orthorotation is technologically executed to twist the ZPE flux into our 3-space.

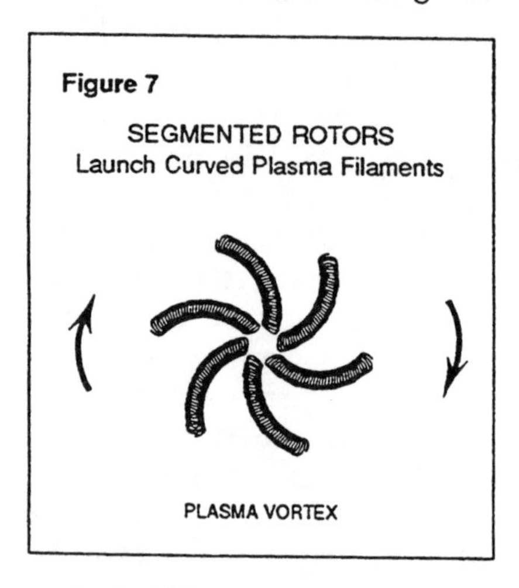

A more direct means to yield a pulsed orthorotation of the ZPE flux may arise from abruptly pulsed, opposing electromagnetic fields. When EM fields buck, there is no net field vector, yet the stress-energy tensor and electrical potential abruptly change in the region of space subject to the bucking fields. This in turn influences the action of the ZPE. The abrupt pressure and release on the orthogonal ZPE flux can result in an excess orthorotation into our 3-space (Figure 8) if there are nuclei in either a crystal lattice or plasma available to guide the flux along their metric curving, stable, vacuum polarization channels.

A device suggested to accomplish this is the caduceus coil.[37,38] This coil is comprised of two identical windings of opposite helicity. The windings must exhibit identical, mir-

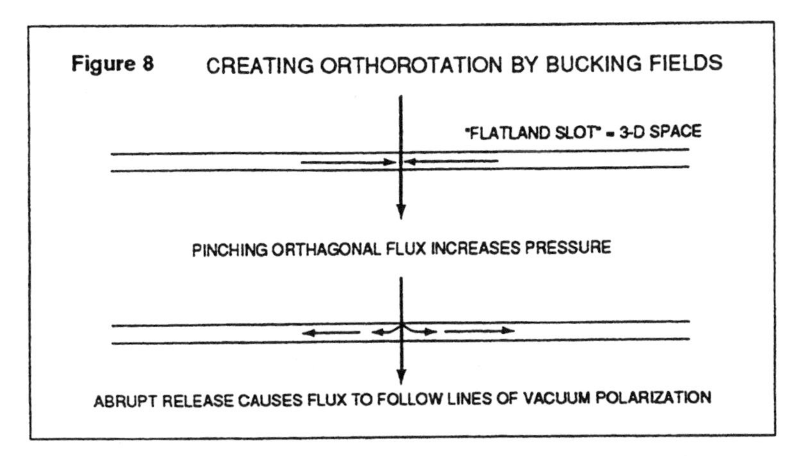

Figure 8 CREATING ORTHOROTATION BY BUCKING FIELDS

ror image symmetry, for a pulse traveling up both windings should have its rising and falling waveform edges aligned. This maximizes the time derivative of the bucking field transient.[39] It has been previously proposed that the rising and falling edges in the bucking field transient produce hyperspatial, toroidal vortex rings that exhibit electromagnetic scalar and longitudinal components in their 3-space projection.[40] Such a form may be directly orthorotating the ZPE yielding time and gravitational anomalies.

It is interesting to note that many inventions which manifest gravitational or energy anomalies utilize the bucking field motif. The rotors of Searl and Carr's disks, Gray's motor, and the Swiss ML converter are propelled by bucking, pulsed electromagnets. The plasma discharges of these devices are directly subjected to these pulsed bucking fields. Tesla repeatedly produced ball lightning in his magnifying transmitter whenever a phasing error condition would have the forward going pulse identically oppose the reflected pulse.[41] In Newman's motor, the coil is so large that the commutator can launch an opposing pulse into it before the previous one travels its length.[42] This also produces a bucking field condition.

Perhaps the most straightforward way to investigate these methods for cohering the ZPE is to combine the above ideas in one experiment. Subject the ion-acoustic oscillations of a

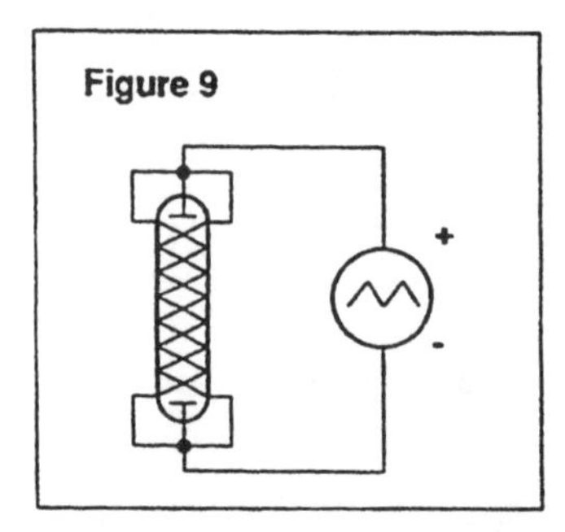

Figure 9

plasma tube to pulsed, electromagnetic opposition by inserting the tube into the core of a caduceus coil (Figure 9). A variable oscillator producing a triangular waveform can be used to trigger the ion-acoustic mode in the tube.[43,44] Coupling the tube's leads directly across the caduceus coil will allow the natural spiking from the ion-acoustic resonance to be the pulsed excitation on the coil itself. A synergistic feedback may be triggered in which the orthorotated ZPE flux couples to the ion-acoustic oscillations further energizing the spiking on the coil. This would yield an increasing macroscopic vacuum polarization as well as gravitational, energy, and time anomalies. A dramatic change in weight of the apparatus would then conclusively demonstrate a zero-point energy coherence.

SUMMARY

Quantum physics has given a novel view of the fabric of space where a hyperspatial electric flux manifests a virtual plasma called the zero-point energy. Theories of system self-organization have shown that in principle, such a plasma can exhibit coherent self-organized modes as observed in plasmas of matter. Quantum electrodynamics has shown how the charged elementary particles are intertwined with the ZPE action and that the nuclei of ions may provide the key for orthorotating the ZPE flux. Thus, a macroscopic vacuum polarization may arise in the self-organized, collective behavior of a plasma's ions. The work of many inventors has shown how ion vortices and vortex rings can yield energy and gravitational anomalies. Such anomalies can be further magnified by subjecting the plasma to abruptly pulsed, bucking electromagnetic fields. This suggests an experiment where a pulsed caduceus coil further excites the ion-acoustic oscillations in a plasma tube inserted through its core. An exhibited gravitational anomaly by a change in weight of the apparatus, or altering the frequency of a nearby mechanical oscillator, would demonstrate a ZPE coherence, for the ZPE is the only energy appreciable enough to cause a space-time metric distortion by technological means. It is hoped that many investigators repeat this suggested experiment and share their results, for only a repeating experiment can shift the paradigm in science to allow the worldwide engineering and acceptance of a wondrous, new technology.

REFERENCES

1. G. Zukav, *Dancing Wu Li Masters*, Bantam Books, NY, 1980.
   This overview of modern physics includes an excellent discussion of nonlocal phenomena arising in quantum mechanics.
2. A. Shimony, "The Reality of the Quantum World," *Sci. Amr.* 258 (1), 46 (January, 1988).
   This article overviews the experiments that demonstrate the nonlocal nature of quantum events. See also reference 46.
3. P. B. Burt, *Quantum Mechanics and Nonlinear Waves*, Harwood Academic, NY 1981.
   This text promotes the view that all quantum mechanical particles and systems arise from a persistent, nonlinear self-interaction with the ZPE.
4. T. H. Boyer, "Random Electrodynamics: The Theory of Classical Electrodynamics with Classical Electromagnetic Zero-Point Radiation," *Phys. Rev.* D, 11 (4), 790-808 (1975).
   The author shows that quantum effects arise because of matter's interaction with the ZPE.
5. H. E. Puthoff, "Ground State of Hydrogen as a Zero-Point Fluctuation Determined State," *Phys: Rev.* D, 35 (10), 3266 (1987).
   The author shows that the ground state of the hydrogen atom results from an equilibrium between radiation emitted due to acceleration of the electron and radiation absorbed from the zero-point fluctuations.

6. A. Hasegawa, "Self-Organization Processes in Continuous Media," *Adv. Phys.* 34 (1), 1-42 (1985).
This paper reviews the behavior of nonlinear, dissipative continuous media. Such media can exhibit the formation of ordered structures even when starting from an initially turbulent state. Examples include magnetohydrodynamic fluids, magnetized plasmas, and atmospheres of rotating planets. The relationship between the onset of chaos and self-organization in a soliton system as well as vortex solutions is also discussed.

7. M. B. King, "Cohering the Zero-Point Energy," Proceedings of the 1986 International Tesla Symposium, pp. 4-13, 4-32, International Tesla Society, Colorado Springs, CO, 1986.
This paper shows that it is theoretically possible to cohere the zero-point energy by merging theories of the ZPE with theories of system self-organization. The key component for interacting with the ZPE is nuclei of a plasma. A hyperspatial flux model of the ZPE is introduced and shown to be affected by abruptly pulsed, opposing electromagnetic fields.

8. H. Haken, *Synergetics,* Springer Verlag, NY, 1971.
This text shows with general system theory mathematics the conditions for self-organization. The formalism can be applied to any system.

9. M. Suzuki, "Fluctuation and Formation of Macroscopic Order in Nonequilibrium Systems," *Prog. Theor. Phy. Suppl.* 79, 125-140 (1984).
The role of fluctuation and nonlinearity in the formation process of macroscopic order is discussed. A coherent interaction model is presented to study self-organizing processes.

10. S. Firrao, "Physical Foundation of Self-Organizing Systems Theory," *Cybernetica* 17 (2), 107-24 (1984).
This paper discusses the contradiction between the law of increased entropy and the fundamental hypothesis of any theory of self-organizing systems.

11. L. Abbott, "The Mystery of the Cosmological Constant," *Sci. Amr.,* 106-112 (May, 1988).
This article discusses an overt philosophical paradox of modern physics: The success of the standard model for elementary particles based on a plenum of zero-point energy versus the amount of energy actually observed.

12. O. H. Abroskina, G. Kh. Kitaeva, A. N. Penin, "The Effective Brightness of Zero-Point Fluctuations of the Electromagnetic Vacuum of Parametric Scattering of Light," *Sov. Phys. Dokl.*, 30 (1), 67 (1985).
This paper describes an experiment that shows the effective brightness of the zero-point fluctuations is independent of the presence of reflectors and absorbers.

13. H. Everett, "Relative State Formulation of Quantum Mechanics," *Rev. Mod. Phys.* 29 (3), 454 (1957). Also, B.S. DeWitt, N. Graham, *The Many Worlds Interpretation of Quantum Mechanics*, Princeton University Press, 1973.
Everett introduces a self-consistent formulation of quantum mechanics without invoking postulates regarding the observer. The formulation yields a superspace containing an infinite number of three dimensional spaces.

14. P. Davies, *Other Worlds*, Simon and Schuster, NY, 1980.
This book describes the various views of superspace arising in modern physics.

15. J. Gribbon, *In Search of Schrödinger's Cat*, Bantam Books, NY, 1984.
An historic development of quantum mechanics is presented including a thorough explanation of the many worlds interpretation.

16. D. Bohm, *Wholeness and the Implicate Order*, Routledge and Kegan Paul, London, 1980.
Bohm presents his theory where all explicate phenomena have their roots in an implicate order (in the ZPE) where there exists a universal holism. A quantum potential is used to describe the nonlocal connections.

See also D. Bohm, F. D. Peat, *Science, Order, and Creativity*, Bantam Books, NY, 1987.

17. O. Klein, "The Atomicity of Electricity as a Quantum Theory Law," *Nature* 118, 516 (1926).
In support of Kaluza's five dimensional unified field theory, Klein suggests that the origin of Planck's quantum may be due to a periodicity in a higher physical dimension. The small value of the characteristic length in this dimension is related to Planck's constant.

18. J. A. Wheeler, *Geometrodynamics*, Academic Press, NY, 1962.
Wheeler derives the modern view of the fabric of space by

applying general relativity to the zero-point energy of quantum mechanics. The result is a dynamic maelstrom of nonlocal, microscopic channels across superspace called "wormholes" whose diameters are on the order of 10^{-33} cm. An electric flux passes through these channels with an energy density on the order of 10^{93} grams/cm^3.

19. B. Toben, F. Wolf, *Space-Time and Beyond*, E.P. Dutton, NY, 1982.
This book is an illustrated introduction to the theories of superspace, geometrodynamics and the many worlds interpretation of quantum mechanics.

20. J. A. Richards, F. W. Sears, M. R. Wehr, M. W. Zemansky, *Modern University Physics*, Addison Wesley, Reading, Mass. 1960, pp. 767-771.
A lucid derivation of the Lorentz transformation is given, where both the stationary observer and moving observer insist that each remains in the center of the same expanding spherical wavefront of light.

21. W. M. Honig, *The Quantum and Beyond*, Philosophical Library, NY, 1986.
An expanding toroidal structure called the "photex" is presented as a model for light. It is supported in a two fluid ether that resembles a turbulent virtual plasma. From this model, the author derives a unified theory that is consistent with both quantum mechanics and special relativity.

22. R. G. Jahn, B. J. Dunne, *Margins of Reality*, Harcourt, Brace, Jovanovich, NY, 1987.
This text reviews years of experiments where subjects influenced probablistic processes. The experiments involved thousands of trials using electronic noise sources.

23. I. Prigogine, I. Stengles, *Order out of Chaos*, Bantam Books, NY, 1984.
This book describes Prigogine's Nobel Prize winning contribution to the field of system self-organization in thermodynamics.

24. S. I. Putterman, P. H. Roberts, "Random Waves in a Classical Nonlinear Grassman Field," *Physica* 131 A, 51-63 (1985).
This paper shows that Fermi statistics can arise from the particles' nonlinear interaction with the ZPE. The authors mathematically show that the nonlinear Langevin formalism model-

ing the ZPE allows energy to be extracted from certain modes of the ZPE. The authors are critical of this result.

25. M. Kono, E. Miyashita, "Modon Formation in the Nonlinear Development of the Collisional Drift Wave Instability," *Phys. Fluids,* 31 (2), 326-331 (1988).
    Nonlinear simulations of plasmas show coherent structures are formed from a turbulent state. A process of small vortices fusing into larger vortices is described.
26. R. Horiuchi, T. Sato, "Three-dimensional Self-Organization of a Magnetohydrodynamic Plasma," *Phys. Rev. Lett.* 55 (2), 211-213 (1985).
    This simulation demonstrates self-organizational processes occurring in a plasma.
27. W. H. Bostick, "Experimental Study of Plasmoids," *Phys. Rev.* 106 (3), 404 (1957).
    Vortex ring structures are experimentally observed in a plasma including plasmoid pair production. A "quantum condition" for plasmoid stability is identified in the ratio of the toroidal to poloidal diameters.
28. P. O. Johnson, "Ball Lightning and Self-Containing Electro-Magnetic Fields," *Am. J. Phys.,* 33, 119 (1965).
    A vortex ring model for ball lightning is presented.
29. E. J. Lerner, "The Big Bang Never Happened," *Discover* 9 (6), 70 (June 1988).
    This article overviews Hannes Alven's theory of a plasma self-organizing universe. Plasma filaments and vortices play a key role in galaxy formation.
30. I. R. Senitzky, "Radiation-Reaction and Vacuum Field Effects in Heisenberg-Picture Quantum Electrodynamics," *Phys. Rev. Lett.* 31 (15), 955 (1973).
    This paper shows all elementary particles are inseparably intertwined with the ZPE and this interaction is the basis of a charge particle's radiation characteristics.
31. M. B. King, "Macroscopic Vacuum Polarization," Proceedings of the Tesla Centennial Symposium, International Tesla Society, Colorado Springs, pp. 99-107, 1984.
    It is speculated that the ion-acoustic mode of a plasma launches and detects macroscopic vacuum polarized modes in the ZPE. Numerous references on ion-acoustic plasma behavior are cited.

32. T. H. Moray, J. E. Moray, *The Sea of Energy*, Cosray Research Institute, Salt Lake City, 1978.
The history of T. H. Moray's radiant energy invention is presented with many testimonials from witnesses. Moray utilized ion oscillations in his tubes.

33. D. A. Kelly, *The Manual of Free Energy Devices and Systems*, Cadake Industries, P.O. Box 1866 Clayton, GA, 30525.
This book contains descriptions and references to many devices associated with claims of free energy or gravitational anomalies. Included are the inventions of Searl, Carr, Gray, Newman, and the Swiss ML Converter.

34. B. E. DePalma, E. C. Edwards, "The Force Machine Experiments," 1973.

This paper describes DePalma's experiments in forced mechanical precession where inertial anomalies were observed. The "force machine" contained two counter-spinning flywheels mounted in a cylinder that slowly rotated to force the flywheels into precession. The entire apparatus was mounted on a large pendulum. Under forced precession of the flywheels, the period of the pendulum altered.

35. S. Seike, *The Principles of Ultrarelativity*, G-Research Laboratory, Tokyo, Japan, 1978.
Seike proposes the existence of a physical hyperspace with an electrical energy flux that flows orthogonally to our three dimensional space. To orthorotate this flux into 3-space requires a four dimensional rotation. Seike calculates how the 3-space projection of the four dimensional rotation would appear in our 3-space. Then by actualizing the dynamics of this 3-space projection with the motion of charge, the hyperspatial rotation is naturally induced.

36. C. W. Cho, "Tetrahedral Physics," 449 Izumi, Komae City, Tokyo, Japan, 1971.
Cho describes in detail one of Seike's (reference 35) hyperspatial, four dimensional rotating forms called the "resonating electromagnetic field" (RMF). The RMF is generated by rapidly switching electric charge in a specific way among four spheres located at the vertices of a tetrahedron. The dynamics of the switching produce two orthogonal rotational modes, one precessing relative to the other. The hyperspatial rotation induced by this 3-space projection is described as a "dynamical

Klein bottle." It is claimed that this experiment will produce inertial and gravitational anomalies.

37. W. B. Smith, *The New Science,* Fern-Graphic Publ., Mississauga, Ontario, 1964.
This esoteric work claims the energized caduceus coil (opposing helical windings on a ferrite core) creates a "tempic field" (an alteration in the time component of the space-time metric).

38. G. Burridge, "The Smith Coil," *Psychic Observer,* 35 (5), 410-16 (1979).
This article explains how to wind a caduceus coil and reports on some observations made by investigators experimenting with this coil. See also reference 47.

39. T. E. Bearden, *Fer-De-Lance: A Briefing on Soviet Scalar Electromagnetic Weapons,* Tesla Book Co., Millbrae, CA ,1986.
The author claims the Soviet Union has developed scalar electromagnetic weapons and offers evidence of their testing. Note 62 (pp. 107-108) stresses the importance of the rise time on the pulse creating the opposing electromagnetic fields and shows the coherent ZPE energy content is proportional to the square of the time derivative of the pulse.

40. M. B. King, "Cohering the Zero-Point Energy," (reference 7).
A hyperspatial model for a "columnar standing wave" or "scalar wave" is proposed consisting of toroidal rotations of electric flux in the fourth dimension. The projection of this form into 3-space yields scalar and longitudinal electromagnetic components. No energy propagates in 3-space through these components, but energy may propagate in the parallel hyperspace. The abruptly pulsed caduceus coil launches such a structure.

41. H. W. Secor, "The Tesla High Frequency Oscillator," *Electrical Experimenter* 3, 615 (1916).
This article describes Tesla's magnifying transmitter and the bucking pulse accidents the produced ball lightning.

42. J. W. Newman, *The Energy Machine of Joseph Newman,* Joseph Newman Publ. Co., Lucedale, Miss., 1984.
The author describes his theory and his energy producing invention consisting of a large coil, rotating magnet, battery and commutator which directs electrical pulses through the coil.

43. G. Obelensky, private communication, 1978.

44. V. Hart, private communication, 1982.

45. M. Kaku, J. Hainer, *Beyond Einstein: The Cosmic Quest for the Theory of the Universe*, Bantam Books, NY 1987.
This book is a layman's introduction to the various unification theories of modern physics. Superstring theory appears to be the most promising for unifying all four fundamental forces (strong nuclear, weak nuclear, electromagnetic and gravitation). The fundamental unit in this theory is a filament on the order of 10^{-33}cm (same size as Wheeler's quantum wormholes) that can form into a Mobius loop. The theory avoids the problems of renormalization by not using point particles.

46. A. Datta, D. Home, A. Raychaudhuri, "A Curious Gedanken Example of the Einstein-Podolsky-Rosen Paradox Using CP Nonconservation," *Phys. Lett.* 123 (1), 4 (1987).
This paper discusses a version of the EPR paradox which shows that a change in kaon emission rate on one side of an apparatus producing kaon pairs can be induced by altering the kaon flight on the other side. This effect occurs in a spacelike or superluminal way. If this can be experimentally realized, the EPR effect would allow sending a signal back in time. The resulting time paradox could be averted by accepting the many world's interpretation of quantum mechanics.[13] For a layman's discussion see J. G. Cramer, "Paradoxes and FTL Communication," *Analog*, 122 (July 1988).

47. E. Dollard, "Van Tassel's Caduceus Coils," private communication, 1988.
Van Tassel experimented with numerous caduceus coils that often contained quartz crystal cores. He stressed that the crossover angle for the two opposing windings should be 22.5 degrees.

ELECTROLYTIC FUSION: A ZERO-POINT ENERGY COHERENCE?

June 1989

ABSTRACT

A zero-point energy hypothesis is proposed to explain the anomalous heat in the Pons/Fleischmann cold fusion experiment. It is suggested that a coherent, collective proton (deuteron) resonance occurs in the supersaturated plasma within the experiment's palladium electrode that creates a macroscopic vacuum polarization. This coherence is optimized by casting the palladium electrode as a pure crystal, and treating its surface to allow maximal hydrogen (deuterium) adsorption. Imposing abruptly bucking electromagnetic fields on the proton (deuteron) plasma during resonance may enhance the zero-point energy interaction sufficiently to yield a measurable gravitational anomaly. Electricity could be tapped directly by a circuit in series across the experiment's palladium rod.

INTRODUCTION

The Pons/Fleischmann "cold fusion" experiment[1] has surprised the scientific world with claims of tremendous heat production without the corresponding quantities of fusion by-products such as neutrons, tritium and helium. Currently Pons claims a heat output exceeding power input by a ratio of one hundred to one.[2] In an earlier trial the heat was so powerful that it vaporized the experiment's palladium electrode.[3] Furthermore, Pons mentioned he produced anomalous heat in another experiment using light water (H_2O) instead of heavy water (D_2O).[4] The scientific community cannot explain the origin of this anomalous heat.[5] Also, it is apparently difficult to replicate this result, for at this time there are only four universities (University of Utah, Texas A&M, Stanford, and Case Western Reserve) who have reported that they successfully produced the anomalous heat.[6] This is poor repeatability for, no doubt, hundreds of attempts at this experiment have been made worldwide. However, these four successful universities have been able to continue trial after trial of replication with a high percentage of success. Clearly they are doing something right.

This paper focuses on that "something" from a novel perspective: The source of the anomalous heat may be neither fusion nor a chemical reaction, but rather, it could arise from a cohering interaction with the zero-point energy (ZPE), the energetic, high frequency, random fluctuations of electric flux imbedded within the fabric of space.[7,8] It is suggested that such a coherence is created by a collective, synchronous oscillation of the deuterium (or hydrogen) nuclei within the experiment's palladium electrode. If so, casting the palladium as a single crystal would increase this coherence.

If the zero-point energy hypothesis is correct, the energy output may be magnified by the use of a pulsed caduceus coil surrounding the palladium electrode. Moreover, a gravitational anomaly could manifest a weight change in the apparatus as well.

FUSION?

Before resorting to the ZPE hypothesis, what evidence is there that fusion is not necessarily the source of anomalous heat in the Pons/Fleischmann experiment? The standard setup is described[1] as follows (Figure 1): A palladium cathode is immersed in an electrolytic solution of 0.1 molar LiOD in 99.5% D_2O + 0.5% H_2O. The lithium deuteroxide (LiOD) is added to make the electrolyte conductive. The palladium cathode is surrounded by a bare platinum wire anode wrapped on a cage of glass rods. The platinum is attached to a positive DC voltage while the palladium is charged negatively. Through electrolysis of the D_2O, the deuterium atoms are absorbed into the palladium while oxygen accumulates at the platinum anode. The deuterium atom ionizes with its electrons entering the band structure of the palladium, and the deuterons settle into the octahedral interstitial sites of the palladium crystal lattice.[9] After a couple of weeks of charging, the palladium rod is supersaturated with deuterons, and it has a crystal lattice structure like NaCl.[9] All lattice sites are occupied, and the excess free deuterons form a "protonic fluid" which can aid electrical conduction.[10] The deuterium density is greater than that of liquid hydrogen.[11] The interstitial lattice sites are shallow potential wells[12] allowing for high deuteron mobility, and presumably, an enhanced probability of fusion events via tunneling through the repulsive, proton Coulomb barrier. The standard fusion reactions are[13]

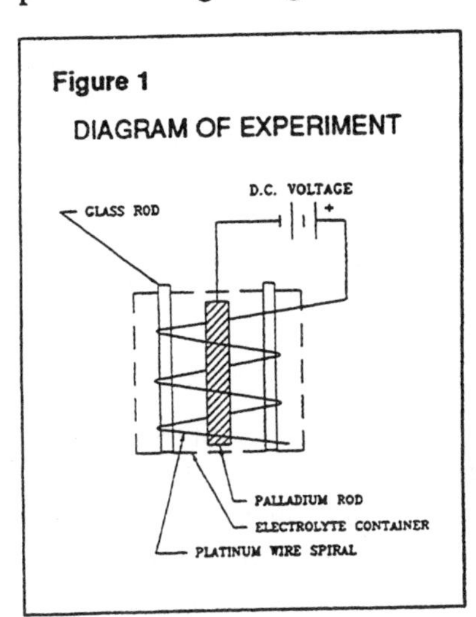

Figure 1
DIAGRAM OF EXPERIMENT

$$D + D \longrightarrow {}^3He + n \quad (4\ MeV)$$

$$D + D \longrightarrow T + p \quad (3.27\ MeV)$$

with equal probability (about 50%). The physics community has been quite vocal arguing that these standard reactions cannot be the source of the anomalous heat, for the corresponding emission of neutrons would have killed the unprotected researchers.[14] Some neutron emissions were detected and confirmed by other investigators,[15] indicating that these reactions are occurring, but at a rate a billion times too slow to account for the anomalous heat production.[1]

Walling and Simmons[16] have proposed the reaction

$$D + D \longrightarrow {}^{4}He + Beta\ (24\ MeV)$$

where Beta is a collective excitation in the palladium electron bands. Along similar lines, Hagelstein[17] has proposed

$$D + D \longrightarrow {}^{4}He + phonon\ (24\ MeV)$$

where the phonon is a coherent vibrational mode in the palladium lattice. The phonon could help trigger deuteron tunneling yielding more fusion events. Another similar reaction would be

$$D + D \longrightarrow {}^{4}He + plasmon\ (24\ MeV)$$

where the plasmon is a collective excitation (soliton) in the deuteron plasma within the supersaturated palladium lattice. If this plasmon propagated like a plasma shock wave,[18] it could then trigger new fusion events through its own compression. So far too little ^{4}He has been detected to prove that these reactions are occurring. It is interesting to note that the Walling and Simmons theory predicts a high rate (600 times greater than D-D fusion) for

$$H + D \longrightarrow {}^{3}He + gamma\ (5.6\ MeV)$$

This suggests that fusion might even be observed using water of natural deuterium content. This is significant for Pons has observed anomalous heat production using ordinary light water, which would imply at first thought that fusion could not be the source of the heat. However, the Walling and Simmons theory may still allow it. The proof awaits the detection of helium which so far has not been observed in sufficient abundance to account for the heat.

To the orthodox scientific community, the heat production in the light water experiment would imply that a chemical reaction of some type was the explanation for the energy.

Pauling has suggested that heat is released in disordering the palladium lattice,[19] while a German team has suggested that hydrogen ignition is occurring at the air-water interface.[20] In fact the experimental apparatus could explode by the following runaway reaction: If the electrical potential on a supersaturated palladium rod is abruptly released, hydrogen would immediately evolve from the rod. If the rod were exposed to the air and this hydrogen ignited, it would further heat the rod releasing more hydrogen. This could explain some of the explosive accidents, but the chemical reactions do not yield sufficient energy to account for all the observations. Pons has observed heat production in excess of 4 million joules per cubic centimeter of electrode volume for experimental times in excess of 120 hours in his early experiments,[1] and he has recently claimed a hundred times more heat power output than power input.[2] No chemical explanation has been proposed to account for this long persistence of heat generation.

As Pons and Fleischmann keep improving their efficiency in anomalous heat production, and if sufficient helium or tritium still remain undetected, the scientific community would be confronted with an energy anomaly of tremendous magnitude. Is there another energy hypothesis available that can be supported with the standard scientific literature?

ZERO-POINT ENERGY

Modern quantum mechanics has come to recognize the existence of an all-pervading energy imbedded within the fabric of space consisting of tremendous, high frequency, random electrical fluctuations called the zero-point energy.[7,8] Zero-point refers to zero degrees Kelvin, meaning these fluctuations are inherent to the vacuum of totally empty space in the absence of all heat, matter and propagating radiation. There is a school of thought, well represented in the physics literature (e.g. Boyer[7]), that treats this energy as physically real, and shows that quantum mechanical effects arise because of matter's intimate interaction with it. It is not a popular view since the nonlinear mathematics required to make quantitative predictions can often become overwhelming. Nonetheless numerous models have been successfully created (e.g. Puthoff's analysis of the hydrogen atom[21]), and the term "vacuum polarization" is extensively used to describe the elementary particles' interaction with the ZPE.

Can this energy be tapped as a source? Today most of the scientific community would answer no, for it would appear to be a violation of thermodynamics' law of entropy where random fluctuations would have to spontaneously self-organize and become coherent. However, there is another school of thought, also well represented in the scientific journals, that deals with the phenomena of system self-organization.[22] Ilya Prigogine[23] won the 1977 Nobel prize in chemistry for identifying the conditions under which a system would self-organize: The system must be nonlinear, far from equilibrium, and have an energy flux through it. These conditions are stated in general systems terms,[24] and the published theories of the ZPE and its interaction with matter can fulfill these conditions under certain circumstances.[25] By combining the theories of the zero-point energy with the theories of system self-organization, a speculative hypothesis for tapping the ZPE as an energy source can be created which does not violate modern physics. It does, however, require an experiment to prove it.

MACROSCOPIC VACUUM POLARIZATION

To induce a cohering interaction with the ZPE requires working with those elementary particles whose vacuum polarization is the most stable and coherent. Quantum electrodynamics shows, in a first order description, that an atom's nucleus induces stable vacuum polarization lines converging toward it,[26] while electrons, especially those in a metal's conduction band have a cloud-like, incoherent interaction with the ZPE.[27] The synchronous, abrupt motion of many nuclei could consequently create macroscopic vacuum polarization effects. This may be supported experimentally with the observations of anomalies associated with the ion-acoustic oscillations of a plasma.[28] In the 1930's an inventor, T. H. Moray,[29] exploited the ion-acoustic anomalies to power his radiant energy device. Moray stressed the importance of ion oscillations in the plasma tubes of his well-witnessed invention, which was reported to produce over 50 kilowatts of electricity—astonishing those scientists who closely examined it. The coherent oscillations or abrupt, synchronous motion of many nuclei (or protons) may be the key for inducing a ZPE coherence.

PROTON LASER

In a gas plasma it is often difficult to excite and maintain coherent ion-acoustic oscillations, for there is turbulence and numerous collisions. Is there a better medium to induce synchronous motion of nuclei? The description of how hydrogen is stored in palladium is that of free protons. The hydrogen's electrons enter the palladium's electron d shell and band structure. The protons tend to settle into shallow potential wells at interstitial lattice sites where they are free to oscillate. These protons interact with their neighbors in what is described as a "soft lattice" that exhibits resonant vibrations at optical frequencies (on the order of 10^{14} Hz).[30] If these protons can be triggered to oscillate synchronously in phase, then a "proton laser" would be created that could exhibit a tremendous ZPE macroscopic vacuum polarization. In neutron scattering experiments on palladium hydride there was observed an anomalous, large amplitude oscillation of protons in the palladium's crystal lattice [100] direction.[30] This suggests that a perfect crystal lattice of palladium hydride would be ideal for supporting coherent proton or deuteron oscillations.

Deuterons exhibit greater stability than protons in their optical oscillations for their greater mass tends to make the oscillation amplitude smaller thus reducing the tendency to slip out of the shallow potential wells.[31] Normally protons exhibit diffusion by hopping to neighboring interstitial sites by thermal activation or tunneling. At room temperatures this diffusion is prevalent and would tend to wash out any phase coherence in the optical oscillations. However, if the palladium were supersaturated with hydrogen or deuterons (as is the case in the Pons/Fleischmann experiment), all the interstitial sites would be occupied limiting the diffusion.[32] Then a maximal number of protons, constrained to their sites by the presence of their neighbors, could exhibit phase coherence in their optical vibrations. Thus a supersaturated palladium crystal becomes an effective medium for the proton laser.

There are other, lower frequency modes that couple to the optical, soft lattice vibrations. The palladium lattice itself supports acoustical phonons that couple to the protons.[30] In addition there is electron band coupling to the protons as well.[33] Of particular interest are low frequency modes involving the whole crystal,[34] for these could allow direct electrical coupling to the energetic, high frequency, ZPE cohering interaction of the soft lattice. Such modes include macroscopic, large wavelength phonons (these could be associated with piezoelectric vibrations), proton currents, group ion-acoustic resonances, solitons,[35] and shock wave formation. The highly nonlinear plasma of the supersaturated proton (or deuteron) "fluid" offers a wealth of opportunities for collective interactions that could couple the high frequency resonances to the lower frequency modes, just as it does in gas plasmas.[36] This interaction could also be used to electrically stimulate the synchronous, soft lattice oscillations as well as couple the resulting ZPE interaction to lower frequencies where the energy could be tapped directly as electricity.

PALLADIUM PREPARATION

In order to optimize the macroscopic coherent effects it is desirable to produce a perfect crystal of supersaturated palladium hydride. Single crystals are normally not produced by standard metallurgical processes such as extrusion or cold casting. In fact, it is generally undesirable to cast single crystals since they are quite fragile when compared to the strength achieved by fine grain casting.[37] Nonetheless the technology for creating single metallic crystals has evolved from the semi-conductor industry where pure crystal silicon and germanium substrates are needed.[38] The process for pulling a crystal is extensively discussed by Paoriei.[39] Figure 2 illustrates an example. The entire casting mold and pure[40,41] liquid metal are immersed in a furnace. At the bottom tip of the casting mold is a water cooled chill block where the solidification will first occur. The mold is then slowly lowered from the furnace, and the crystal will grow at the solid-liquid interface. The process should be done in vacuum or in an inert atmosphere (e.g. argon) in order to avoid contamination.

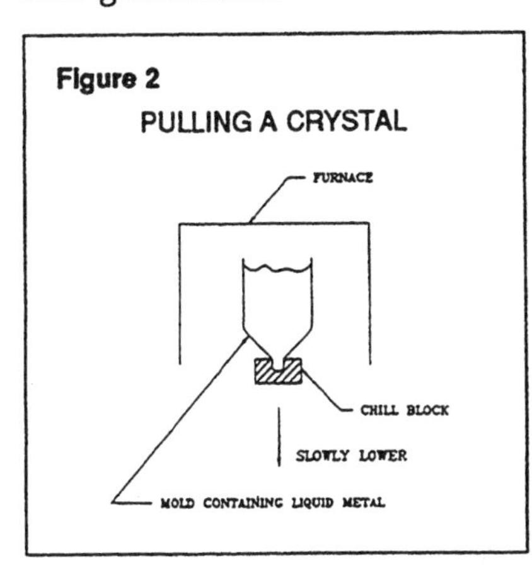

This process will produce a more uniform substrate than cold casting a pure metal where numerous grain boundaries, dislocations and gas gaps can occur[37]. It may well be possible to produce some anomalous heat of a ZPE origin in experiments with cold cast palladium, since columnar grains of single crystals are formed in this process. However, the largest macroscopic resonances would be achieved by the largest crystals of palladium hydride.

The treatment of the palladium rod's surface can have a dominating effect on the palladium's ability to occlude hydrogen. Exposure to the air can result in a blocking oxide layer.[42] Nitrogen, carbon, dust or other impurities can poison the rod's surface as well. Also, the rod cannot be directly handled without introducing contaminants. Many experiments have failed because the palladium's surface was so contaminated that the metal could not adsorb the deuterium. For optimal results the palladium should remain sealed in an inert atmosphere (e.g. argon) from the time it is cast until it is immersed in the experiment's electrolyte.[43]

On the practical side, early investigators found that machining or cold working the metal's surface can aid in hydrogen occlusion.[44] Not only could this remove the surface impurities, but also it would produce a network of stress rifts that would channel the hydrogen into the metal.[45] In fact it was shown that a very smooth surface produced by annealing palladium in vacuum blocked its ability to occlude hydrogen. Occlusion could be recovered by a moderate amount of cold working.[46] However, the problem with cold working is that it will damage the crystal that was so carefully grown.

A less disruptive surface treatment is to coat the palladium crystal with a thin layer of palladium black or other "hydrogen transfer catalyst" such as platinum black, copper powder, or uranium hydride.[47] This is often done to help other metals occlude hydrogen. Palladium black is a coarse, sponge-like form of palladium that offers a large surface area for hydrogen adsorption. Microscopically it appears as a plenum of dendritic microcrystals. Naturally the same care to avoid contaminants must still be maintained before, during, and after the coating process. In this manner a single crystal of palladium can be prepared that will have the ability to absorb large amounts of hydrogen or deuterium. The treatment of the palladium electrode is critical to the success of the Pons/Fleischmann experiment. The difficulty of its proper preparation is probably the reason for the limited repeatability of the heat anomaly.

EXPERIMENTAL APPROACHES

Most investigations of the Pons/Fleischmann experiment have focused on the production of anomalous heat and/or neutrons, but the zero-point energy hypothesis also predicts other effects. These include the possibilities of 1) directly generating electricity with the proton plasma resonances and 2) distorting the space-time metric near the apparatus, yielding a change in gravity or the pace of time.[48] "Scalar" excitation by pulsed, bucking electromagnetic fields could significantly enhance these unusual effects.[49] Since these predictions support the ZPE hypothesis, they will be the main focus here.

All experiments must begin by saturating the palladium rod with deuterium or hydrogen through electrolysis as described in the literature on electrochemistry.[50] Care must be exercised when charging the palladium rod to supersaturation. The charging must be gentle and not too abrupt. The palladium lattice expands as it undergoes transition from the alpha phase to the beta phase.[51] Abrupt changes in the charging current could create dislocations. Hydrogen tends to gather in dislocations and can cause embrittlement of the palladium.[52] Gently heating the solution can help to saturate the rod more quickly. It is important to keep the rod immersed under the electrolytic solution at all times and not expose it to air or the oxygen liberated at the anode. After the rod is supersaturated, it is also important not to abruptly release the charging current for the hydrogen (or deuterium) will evolve from the rod. The experiment should be designed to insure that any leaking hydrogen cannot interact with air or the anode's oxygen. The following electrical excitation experiments will no doubt result in some hydrogen evolution, and if this hydrogen ignited with oxygen, it could explode.

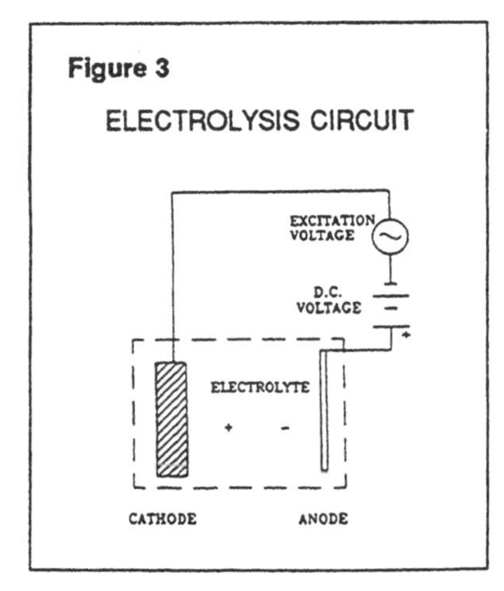

Two types of electrical excitation can be used to trigger the proposed collective proton resonances: A sharp pulse across the palladium rod could induce a direct "ringing" of the resonant modes, or a sweep generator synchronized with a spectrum analyzer could locate the resonant frequencies.

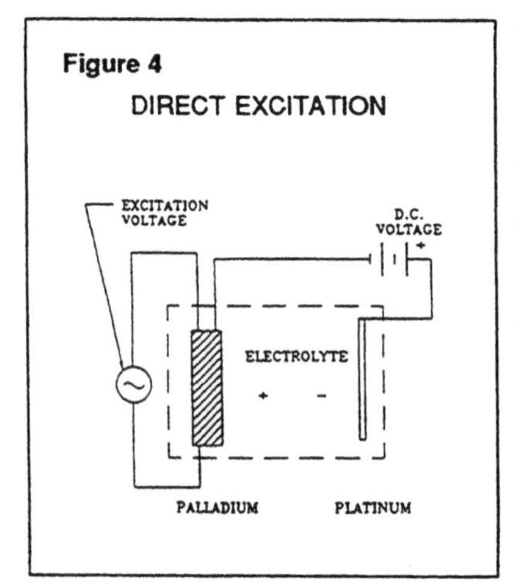

Figure 4
DIRECT EXCITATION

The electrolysis circuit (Figure 3) or a separate excitation circuit (Figure 4) can be used to excite the palladium proton (deuteron) plasma. The electrolysis circuit will only be responsive to low frequency modes because of the series electrolyte. It can be used to induce a "pressure" modulation on the palladium rod as a whole, since the equivalent hydrogen (deuterium) pressure is proportional to the electrolysis voltage.[53] The circuit where the excitation is impressed directly across the palladium rod could be used to not only directly excite the hypothesized proton laser, but also to tap the lower frequency plasma modes directly as electricity. Since high voltage spikes may be produced in the plasma resonances,[54] measurements with oscilloscopes, spectrum analyzers, or transient recorders should use wideband current probes (pickup couplers) around the leads attached to the palladium rod with suitable protection to prevent overloading the instrument. A series capacitive discharge circuit (Figure 5) is recommended for directly pulsing the palladium rod, since it will not be readily damaged by a high voltage response. Such a response could damage a sweep generator or variable oscillator in series with the rod. Experimental protocols similar to those used in plasma

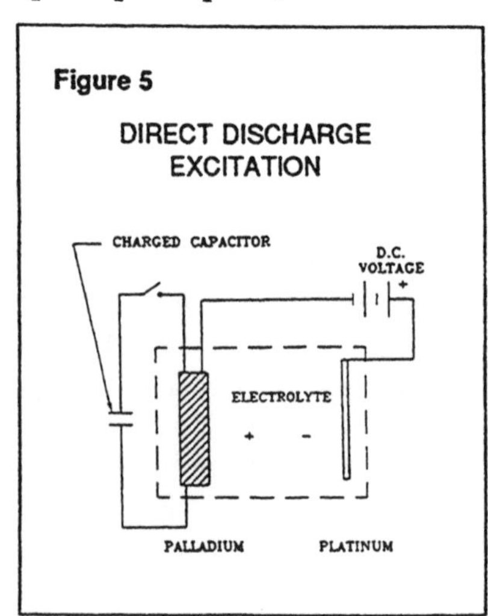

Figure 5
DIRECT DISCHARGE EXCITATION

research can be adapted to explore the wealth of possible resonances of the proton (deuteron) plasma within the supersaturated palladium rod.

If these resonances interact coherently with the ZPE, the effects might be dramatically increased by the use of a pulsed caduceus coil surrounding the palladium rod. A caduceus coil[55] is actually two separate coils wound with perfect mirror image helical symmetry on a hollow tube. The windings cross each other on opposite sides of the tube. The coil is designed to impress abruptly bucking electromagnetic fields within its interior. These opposing fields then induce an abrupt stress on the fabric of space which may influence the vacuum polarization activity of the ZPE as it interacts with the resonating proton (deuteron) plasma. This could yield a greater energy output.

Perhaps the most unusual prediction of the ZPE hypothesis is the possible creation of a gravitational anomaly. This arises from the intimate association between the zero-point energy activity and the space-time metric.[56] This association is explored in quantum gravity theories.[57] Recently Puthoff[58] proposed that gravity actually arises from the action of the ZPE. A weight change in the apparatus during its energetic resonance could be the strongest and most dramatic supporting evidence for the ZPE hypothesis. If the space-time metric is altered, then the pace of time could shift in the vicinity of the apparatus. This could be explored with a mechanical oscillator (e.g. a stop watch or tuning fork) whose frequency would change when brought near the apparatus. (Since electrical oscillators could be influenced by the stray electromagnetic fields from the apparatus, they would not constitute definitive proof of a time change). Also the period of a pendulum could change when brought close to the electrolytic cell. A gravitational effect or a time anomaly is a unique prediction of the zero-point energy hypothesis, and could be used to separate it from the other possible theories to explain the anomalous energy in the Pons/Fleischmann experiment.

SUMMARY

At this time none of the proposed scientific theories have adequately explained the source of anomalous heat in the Pons/Fleischmann experiment. Some novel fusion hypotheses have been suggested, but these await the detection of sufficient helium for their confirmation. The chemical explanations cannot account for the long persistence of heat generation, especially in view of Pons and Fleischmann's continuing improvement of their experiment. The zero-point energy hypothesis is speculative and most unorthodox, but nonetheless it can be supported by the modern physics literature. This hypothesis can account for large amounts of energy and predicts a gravitational or time anomaly that can be used for its confirmation.

The ZPE hypothesis suggests that the optimal energy production will occur when the proton or deuteron plasma within the palladium rod undergoes a coherent, high frequency resonance. To achieve this resonance across the entire palladium rod requires casting it as a pure, single crystal. Furthermore, great care must be taken to insure the surface is properly prepared and free of contaminants so that the palladium electrode would occlude hydrogen or deuterium to supersaturation during electrolysis. The supersaturated state traps the protons (deuterons) in their interstitial lattice sites where they can collectively undergo a phase synchronous, soft lattice oscillation resulting in the hypothesized proton laser. This soft lattice oscillation could then exhibit a macroscopic vacuum polarization in its cohering ZPE interaction, an effect that may be dramatically increased by impressing abruptly bucking electromagnetic fields on the proton (deuteron) plasma by the use of a pulsed caduceus coil. If the predicted gravitational and time anomalies occur in the Pons/Fleischmann experiment, then a bigger discovery than cold fusion is at hand.

ACKNOWLEDGEMENTS

The author wishes to thank David Faust and Oliver Nichelson for their research assistance.

REFERENCES

1. M. Fleischmann, S. Pons, "Electrochemically Induced Nuclear Fusion of Deuterium," *J. Electroanal. Chem.* 261, 301 (1989).
   The authors describe their electrolytic experiment and results. They recognize more heat is generated than can be accounted for by standard fusion theories or chemical reactions.
2. "U. Hopes Device Will Stifle Fusion's Heat Debate," *Salt Lake Tribune*, May 25, 1989, pg 21A.
   This article describes a new electrochemical cell that will be used to make accurate heat measurements. In their experiments, Pons and Fleischmann have increased the power yield to a ratio of 100 to one.
3. "Fusion Claim Electrifies Scientists," *Sci. News* 135, 196 (April 1, 1989).
   This article reports scientists' initial response to the cold fusion announcement. Fleischmann reported that in an early experiment the bottom half of the palladium electrode vaporized.
4. "Hopes for Nuclear Fusion Continue to Turn Cool," *Nature* 338, 691 (April 27, 1989).
   This article summarizes the cold fusion news through the end of April. Most, but not all, of the independent attempts at replication have failed. At an April 17 press conference Pons mentioned he observed an unexpected production of heat in a cell containing ordinary water.
5. Los Alamos Conference on Cold Fusion, Santa Fe, NM, May 23-25, 1989, report by D. Hansen, S. Hassett, private communication, June 1989.
   This conference included both experimental results and theoretical presentations. Texas A&M and Stanford Universities reported successful heat-producing experiments. A Texas A&M graduate student witnessed that a cell emitted sparks and then exploded, leaving a blacken vessel. R. Huggins of Stanford described the elaborate process by which he prepared his palladium rods.
   The theoretical hypotheses presented were:
   (1) Muon catalyzed fusion
   (2) Microcavity explosions due to gas swelling
   (3) Micro-cracking causing electrostatic voltage and deuteron acceleration
   (4) A "symmetry force" causing "globule" formation

(5) Dendrites on the palladium surface causing an electric field that accelerates deuterons
(6) Work of fracture where the palladium is the fuel
(7) Deuterium oxidation
(8) Exothermic alloying of lithium with palladium

None of the fusion or chemical hypotheses proposed to explain the anomalous heat were deemed acceptable under peer review.

6. "New Fusion Criticism Doesn't Faze U.," *Salt Lake Tribune*, May 2, 1989, pg 1A.
This article reviews the criticism of the Pons/Fleischmann experiment. It also mentions the positive results by the four universities whose experiments generate the excess heat.

7. T. H. Boyer, "Random Electrodynamics: The Theory of Classical Electrodynamics with Classical Electromagnetic Zero-Point Energy," *Phys Rev* D 11(4), 790 (1975).
Boyer shows that quantum effects arise because of matter's interaction with the zero-point energy.

8. J. A. Wheeler, *Geometrodynamics*, Academic Press, NY, 1962.
Wheeler derives the modern view of the fabric of space by applying the formalism of general relativity to the ZPE. Due to the large energy density, the fabric of space pinches into blackhole-whitehole pairs that may connect distant regions nonlocally through "wormholes". The fabric of space appears as a turbulent virtual plasma consisting of particles whose size is on the order of Planck's length 10^{-33} cm. The energy density of electric flux passing through each particle is enormous: 10^{93} grams/cm^3.

9. W. M. Mueller, P. Blackledge, G. G. Libowitz, *Metal Hydrides*, Academic Press, NY, 1968, pp 634-652.
This text contains an extensive overview on palladium hydride.

10. D. P. Smith, Hydrogen in Metals, University of Chicago Press, Chicago, 1948.
This text overviews the metallic hydride research through the late 1940's.
Page 124 notes protonic conduction under supersaturation.

11. E. P. Palmer, "Condensed Matter Catalyzed Fusion in Electrolysis and in the Earth," Brigham Young University Colloquium, April 13,1989.
This colloquium briefly described S. E. Jones', et al. cold fusion experiments (reference 15). Palmer mentioned that saturated palladium can have a hydrogen density greater than liquid hydrogen.

12. G. Bambakidis, *Metal Hydrides,* Plenum Press, NY 1981.
This book contains a collection of research papers on metal hydrides.
E. N. Economov, "Superconductivity in Palladium Based Hydrides," pg 3.
This paper notes that the hydrogen soft lattice arises from shallow potential wells. The deuteron sublattice is softer than the hydrogen sublattice due to anharmonic effects and this accounts for a low temperature inverse isotope effect.

13. S. P. Parker, *Nuclear and Particle Physics Source Book,* McGraw Hill, NY. 1988, pg 134.
The standard fusion reactions require two particles in and two particles out in order to conserve both momentum and energy. The symbols used are:

H = hydrogen	^{3}He = isotope helium 3
D = deuterium	^{4}He = isotope helium 4
T = tritium	gamma = high energy photon
n = neutron	p = proton

14. H. W. Lewis,"U. of U.'s Breakthrough Lacks Necessary Conditions of Fusion," *Salt Lake Tribune* Editorial, April 9, 1989, pg 19A.
The main point of this criticism is that the experiment lacks sufficient neutrons for fusion to be the source of heat.

15. S. E. Jones, et. al, "Observation of Cold Nuclear Fusion in Condensed Matter," *Nature* 338, 737 (April 1989).
This paper describes electrolytic fusion experiments using a variety of cathodes including titanium and palladium, a gold anode, and an electrolyte containing a mixture of various metal salts. The experiment focused on neutron counts using a state-of-the-art neutron spectrometer.

16. C. Walling, J. Simmons, "Two Innocent Chemists Look at Cold Fusion," University of Utah, submitted to *J. Phys. Chem.*, May 1989.
The authors propose a theory where the electrons and the palladium lattice form a collective quasi-particle that not only screens deuteron repulsion, but also absorbs the fusion energy though internal conversion. This allows deuterium to fuse directly to helium 4 without neutron production.

17. "MIT Scientists Theorizes on Cold Fusion," *Salt Lake Tribune*, April 13, 1989, pg 2A; also April 15, 1989, pg 2B; also Newsweek, May 8, 1989, pp 48-54.
Hagelstein announces a cold fusion theory that utilizes the collective modes and coherent quantum effects of the deuteron palladium lattice to sustain the fusion reaction.

18. F. F. Chen, *Introduction to Plasma Physics*, Plenum Press, NY, 1977, pg 249.
An introduction to nonlinear modes of a plasma is presented including a clear description of ion-acoustic shock waves.

19. M. D. Lemonick, "Fusion Illusion?" *Time*, May 8, 1989, pp 72-79.
This article overviews the cold fusion discovery. It mentions Dr. Linus Pauling's hypothesis that under high concentrations of deuterium, the palladium lattice may become unstable and deteriorate.

20. G. Schwartz, sci.physics network, April 21, 1989.
The German scientists Kreysa, Marx, and Plieth produced a report for the German Chemical Society in which heat from their experiments was produced by exposing saturated palladium hydride to the air where the metal's surface catalyzed deuterium oxidation.

21. H.E. Puthoff, "Ground-State of Hydrogen as a Zero-Point Fluctuation Determined State," *Phys. Rev.* D 35(10), 3266 (1987).
The author shows that the ground state of the hydrogen atom results from an equilibrium between radiation emitted due to acceleration of the electron and radiation absorbed from the ZPE.

22. S. Firrao, "Physical Foundation of Self-Organizing Systems Theory," *Cybernetica* 17(2), 107, (1984).
This paper discusses the contradiction between the law of entropy and the fundamental hypothesis of any theory of self-organizing systems.
23. I. Prigogine, I. Stengers, *Order Out of Chaos,* Bantam Books, NY, 1984.
This book is a layman's description of Prigogine's Nobel Prize winning contribution to the field of self-organization in thermodynamics.
24. H. Haken, *Synergetics,* Springer-Verlag, NY, 1971.
This text identifies via system theory mathematics the conditions for self-organization. The formalism can then be applied to any system.
25. M. B. King, "Cohering the Zero-Point Energy," Proceedings of the 1986 International Tesla Symposium, International Tesla Society, Colorado Springs 1986, section 4, pp 13-32.
This paper shows that it is theoretically possible to cohere the zero-point energy by merging theories of the ZPE with theories of system self-organization. The key component for interacting with the ZPE is nuclei of a plasma. A hyperspatial flux model of the ZPE is introduced and shown to be affected by abruptly pulsed, opposing electromagnetic fields.
26. F. Scheck, *Leptons, Hadrons and Nuclei,* North Holland Physics Publ., NY 1983, pp 212-223.
This text describes the vacuum polarization of the various elementary particles.
27. I. R. Senitzky, "Radiation Reaction and Vacuum Field Effects in Heisenberg-Picture Quantum Electrodynamics," *Phys. Rev. Lett.* 31(15), 955 (1973).
This paper shows that all elementary particles are inseparably intertwined with the zero-point energy, and this interaction is the basis of a charged particle's radiation characteristics.
28. M. B. King, "Macroscopic Vacuum Polarization," Proceedings of the Tesla Centennial Symposium, International Tesla Society, Colorado Springs, 1984, pp 99-107.
This paper suggests that the ion-acoustic mode of a plasma launches and detects macroscopic vacuum polarized modes in the ZPE. Numerous references on ion-acoustic plasma behavior are cited.

29. T. H. Moray, J. E. Moray, *The Sea of Energy*, Cosray Research Institute, Salt Lake City, 1978.
The history and description of T. Henry Moray's radiant energy invention is presented with numerous testimonies from witnesses.

30. G. Alefeld, J. Volkl, *Hydrogen in Metals I, Basic Properties*, Springer-Verlag, NY, 1978.
T. Springer, "Investigation of Vibrations in Metal Hydrides by Neutron Spectroscopy," pp 75-100.
This chapter discusses the various vibrational modes of metallic hydrides. These include diffusive motions with characteristic times longer than 10^{-12} seconds, host lattice acoustical vibrations up to 10^{13} Hz, and optical phonons of the proton soft lattice in the region of 10^{14} Hz. Palladium hydride exhibits an anomalous anisotropy of large amplitude optical vibrations in the crystal lattice [100] direction.

31. G. Alefeld, J. Volkl, *Hydrogen in Metals II, Application Oriented Properties*, Springer-Verlag, NY, 1978.
E. Wicke, H. Brodowsky, "Hydrogen in Palladium and Palladium Alloys," pp 73-155.
This chapter overviews the palladium hydride research through 1976. There is an extensive discussion of the hydrogen and deuterium oscillations in the lattice, the different properties of these different isotopes, bulk and surface properties, hydrogen mobility, and surface treatments to aid adsorption.

32. R. C. Bowman, "Hydrogen Mobility at High Concentrations," reference 12, pp 109-144.
This paper overviews hydrogen mobility for various metallic hydrides. For palladium hydride PdH_x (where x is the hydrogen concentration) the diffusion rate is proportional to the site blocking factor (1-x).

33. J. P. Burger, "Electron-Phonon Coupling and Superconductivity in Palladium Hydrides and Deuterides," reference 12, pp 243-253.
This paper shows how electron phonon coupling in palladium hydrides is experimentally investigated. It also explains the inverse isotope effect where the superconductivity critical temperature of the deuteride is higher than the hydride. The experiments show that the electrons couple mostly to the optical phonons.

34. H. Wagner, "Elastic Interaction and Phase Transition in Coherent Metal-Hydrogen Alloys," reference 30, pp 5-51.
This chapter discusses metallic lattice vibration modes. These include bulk modes associated with long wavelength phonons, and macroscopic modes which depend on the shape of the material. Different crystal lattice directions exhibit different vibrational characteristics.

35. A. C. Scott, F. Chu, D. W. McLaughlin, "The Soliton: A New Concept in Applied Science," *Proc. IEEE* 61(10), 1443 (1973).
This paper overviews solitary wave theory and shows how solitons (collective modes that do not disperse) arise in nonlinear systems.

36. A. G. Sitenko, *Fluctuations and Nonlinear Wave Interactions in Plasmas*, Pergamon Press, NY, 1982.
This monograph thoroughly overviews the various wave interactions in a plasma. Under non-equilibrium conditions high frequency Langmuir (electron) waves can transfer their energy to longitudinal ion-acoustic waves which may then grow enormously.

37. W. O. Alexander, G. J. Davies, K. A. Reynolds, E. J. Bradbury, *Essential Metallurgy for Engineers*, Van Nostrand Reinhold, Berkshire, England, 1985.
This monograph introduces the various processes for metal purification and processing. Topics include casting, cold working, annealing, grain control, metallic crystal dendrite growth, and defects. It discusses six principle casting defects: Blowholes, cold-shuts, contraction cracks, flash formations, oxide and dross inclusions, and shrinkage cavities.

38. M. F. Ashley, D. R. H. Jones, *Engineering Materials 2, An Introduction to Microstructures*, Processing and Design, Pergamon Press, NY, 1986.
This book is a thorough introductory overview of metallurgical processes. It contains a discussion of coherent crystals, grains, defects, phase boundaries, zone refining purification techniques, cold casting, and crystal casting.

39. K. Lal, *Synthesis, Crystal Growth and Characterization*, North Holland Publ. Co., NY, 1982.
C. Paroriei, "Fundamental Aspects and Techniques of Crystal Growth from the Melt," pg 135.
This chapter thoroughly discusses the various techniques for creating a metallic crystal.

40. G. Foo,"A Critical Analysis of the processing Parameters in Palladium Refining" in M.I. El Guindy, *Precious Metals 1982,* Pergamon Press, NY, 1983, pg 463.
This paper overviews the process for extracting pure palladium from its ore.

41. Dr. Robert Huggens of Stanford University stated at the Los Alamos conference (reference 5) that in order to get sufficiently pure palladium to produce his experiment's heat anomaly, he melted it twelve times under a flushing argon atmosphere, avoided carbon casting crucibles, and removed blocking impurities from the rod's surface.
Reference 31, pg. 58, discusses other purification techniques such as zone melting and electrotransport.Reference 38, pp 34-38, clearly describes zone refining where a slowly moving portion of the metallic rod is melted (in a cast) and allowed to recrystallize. The recrystallization process pushes the impurities into the liquid zone.

42. Reference 10, pg 76, notes an oxide layer can block hydrogen penetration into the metal.
Reference 31, pg 78, notes a smooth, bright, bulk palladium surface contains impurities which impede adsorption.

43. G. Balding, private communication, June 1989.

44. Reference 10, pg 26, notes that moderately cold worked palladium increases its occlusive capacity, but if overworked, it can decrease it.

45. Reference 10, pg 113, notes that metal deformation aids passage of the hydrogen through a network of intragranular rifts which helps occlusion.

46. Reference 10, pg 237, notes that annealing palladium in vacuum can make it inert to hydrogen, but subsequent cold working will cause it to recover.

47. Reference 31, pp 78-88, pp 139-150, discusses surface treatments by "hydrogen transfer catalysts," and describes the activation by palladium black.

48. N. A. Kozyrev, "Possibility of Experimental Study of the Properties of Time," Joint Publication Research Service, Arlington VA, 1968.
The author discusses a group of experiments where vibrating or accelerating gyroscopes are used to alter the pace of time near them.

49. M.B. King, "Demonstrating a Zero-Point Energy Coherence," Proceedings of the 1988 Tesla Symposium, International Tesla Society, Colorado Springs, (to be published).
This paper overviews zero-point energy theories, especially in its relation to gravity. A hyperspatial flux model for the ZPE (reference 25) is reviewed, and it is shown this flux would be influenced by abruptly pulsed, opposing electromagnetic fields. This could result in twisting the ZPE flux into three dimensional space resulting in a space-time metric alteration. A weight change in the apparatus or a change in the pace of time would then imply that a ZPE coherence is occurring. The suggested experiment uses a pulsed, caduceus wound coil surrounding a plasma tube in ion-acoustic resonance.

50. F. A. Lewis, *The Palladium Hydrogen System*, Academic Press, NY, 1967.
This book contains a discussion of electrochemical charging of palladium hydride.

51. Reference 9, pg. 640, notes that the palladium lattice can expand by over 3% in its hydride phase transition from alpha to beta.

52. Reference 31, pg. 305, notes that hydrogen clusters form around dislocations, defects and impurities, and this causes embrittlement of the metal.

53. Reference 1 states an equivalent pressure of 10^{26} atmospheres is occurring in the experiment.
Reference 10, pg. 147, shows that astonishing effective pressures can be achieved by cathodic occlusion of hydrogen.

54. Yu. G. Kalinin, et al., "Observation of Plasma Noise During Turbulent Heating," *Sov. Phys. Dokl.* 14(11), 1074 (1970).
High voltage spikes are observed during plasma turbulence.

55. Reference 25 contains an extensive discussion of the caduceus coil.

56. M. B. King, "Is Artificial Gravity Possible?," University of Pennsylvania, May 1976.
This paper explores the linkage between the zero-point energy and the space-time metric. If the ZPE could be cohered, then gravitational propulsion would be possible. The capacitor thrust experiments of T. Townsend Brown are reviewed as possible evidence for a gravitational effect.

57. N. D. Birrel, P. C. W. Davies, *Quantum Fields in Curved Space,* Cambridge University Press, NY, 1982.
This text overviews quantum gravity theories where the zero-point energy plays a crucial role.

58. H. E. Puthoff, "Gravity as a Zero-Point Fluctuation Force," *Phys. Rev.* A 39(5), 2333 (1989).
This paper shows how gravity arises from an induced effect associated with the zero-point fluctuations of the vacuum. This model of gravity constitutes an "already unified" field theory.

SCALAR CURRENT

October 1989

One of the most unusual claims associated with certain "free energy" devices is the ability to conduct appreciable power on ordinary thin wires without heating them. To an electrical engineer this result is extraordinary and it would constitute a definitive demonstration of a novel form of electromagnetism. Here is suggested an experiment that could produce "cold conduction" and demonstrate an hypothesized phenomenon known as "scalar current."

Scalar current arises by abruptly bucking magnetic fields onto a caduceus wound or a bifilar wound coil (Figure 1). If bucking magnetic fields are impressed onto an ordinary, single wound coil, no current would flow since the magnetic fields cancel. However, impressing these fields onto a caduceus or bifilar coil would allow two oppositely flowing "virtual" currents to occur because, by symmetry of the windings, the opposite current vectors sum to an effective zero current. The currents are described as "virtual" since they are comprised not of electron flow in the wires, but rather a displacement current in the vacuum zero-point energy outside the wire. It is as if the abruptly bucking magnetic fields

ABRUPTLY BUCKING MAGNETIC FIELDS INDUCE SCALAR CURRENT

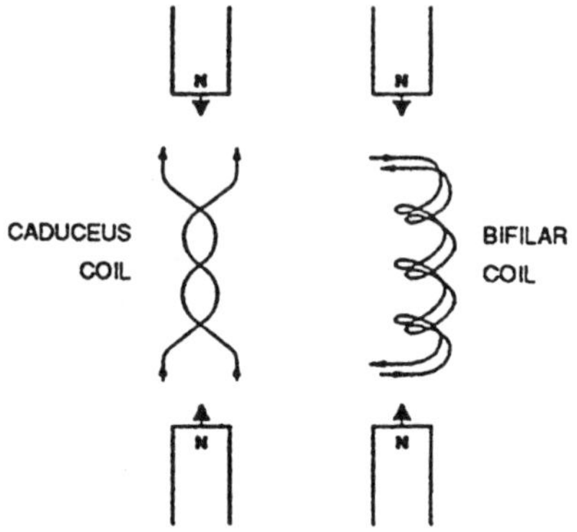

CADUCEUS COIL

BIFILAR COIL

N

N

N

N

FIGURE 1

manifested a pair production of two macroscopic, oppositely rotating, displacement current vortices in the zero-point energy.[1] These vacuum energy vortices are stabilized and supported by the two symmetric windings.

There are many ways to impress the abruptly bucking magnetic fields onto the caduceus or bifilar coil. One method could use two electromagnets with the proper control circuitry to appropriately phase the magnetic fields. In another method the coil could be spun in the air gap between two opposing permanent magnets using brushes and slip rings to tap the scalar current (Figure 2). Bedini described using this method in his "gravity field generator" where he not only reported "cold conduction" but also a weight change in the apparatus as well.[2] A third method could shift the bucking fields onto the coil by either physically oscillating

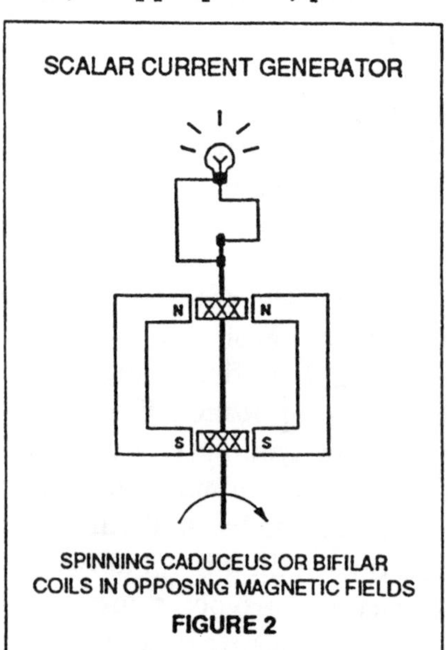

SPINNING CADUCEUS OR BIFILAR COILS IN OPPOSING MAGNETIC FIELDS

FIGURE 2

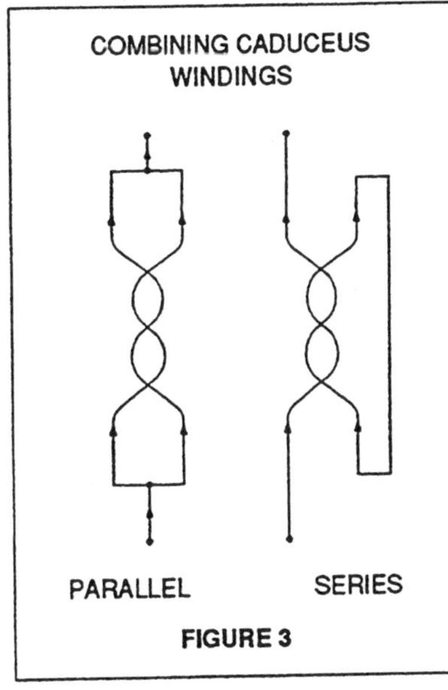

FIGURE 3

opposing permanent magnets or shifting their fields by variable reluctance techniques.[3] Obviously any method that resulted in shifting a bucking magnetic flux onto the caduceus or bifilar coils could be utilized in this experiment.

The experimenter could also explore how to best combine the currents in the opposite windings. The windings could remain separate or be combined in series or parallel (Figures 3 and 4). Another option could add a second caduceus (or bifilar) coil in the air gap at the opposite poles of the bucking magnets so that both ends of the alternating bucking magnets are launching scalar currents (e.g. Figure 2). These two sets of coils could then be combined appropriately to keep the currents in phase to maximize the output.

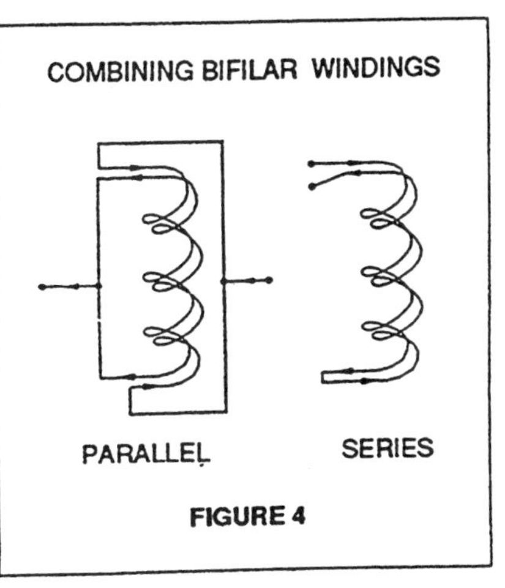

FIGURE 4

An attractive attribute of this suggested experiment is its simplicity. It is hoped that those working with these ideas freely share their results for it will expedite the development of a new technology.

REFERENCES

1. M.B. King, "Cohering the Zero-Point Energy," Proceedings of the 1986 International Tesla Symposium, International Tesla Society, Colorado Springs, 1986, section 4, pp 13-32.

   This paper explains how "free energy" and "antigravity" might be possible with today's physics. By applying the theories of system self-organization to the theories of the zero-point energy a theoretical case is presented with experimental suggestions. Bucking fields and caduceus coils are discussed.

2. T. Bearden, T. Herold, E. Mueller, "Gravity Field Generator Manufactured by John Bedini," Tesla Book Co., Greenville Texas, June 1985. Also J. Bedini, "The Bedini Motor, A Free Energy Device," presented at the Tesla Centennial Symposium, Colorado Springs, August 1984.

   The authors describe their theory and experiments with Bedini's gravity field generator. The device is a Kromrey type generator (U.S. Patent #3,374,376) modified to produce scalar currents.

3. F. Richardson, "Electromagnetic Convertor with Stationary Variable-Reluctance Members," U.S. Patent #4,077,001, 1978.

   This patent describes a method for shifting the magnetic flux of a permanent magnet without moving parts.

AFTERWORD

It turns out that through the years many inventors have wittingly or unwittingly utilized techniques that induced a cohering interaction with the zero-point energy. The study of the plight of these inventors teaches an important lesson: Basically, an invention that violates the scientific paradigm (i.e., the "known" world view) will be rejected or ignored by the scientific community. All patents are disallowed, for they sound like perpetual motion. A working device that can tap appreciable amounts of energy will cause extreme pressure to be brought onto the inventor. He will be squeezed between those who wish to steal it and those who wish to suppress it. A lone inventor has absolutely no chance. Is there anything he can do?

The answer is yes, if the inventor can see the big picture. Primarily, the inventor must understand that humanity rejects any working invention that violates the recognized paradigm or world view. It is nothing personal or novel to our age; this has been the case throughout history. Paradigms have shifted in the history of mankind as well as in the history of science. What causes a paradigm shift is the creation of a repeating experiment. Repetition is most important for if the results are not witnessed by most of the scientific community, the experiment will be ignored. The experiment must also be simple for there is no funding available to produce an experiment that violates the paradigm, at least in the initial stages. But there are young people, students and inventors with that "can do" attitude who would be most willing to repeat an experiment that would uplift the world. These people are our greatest resource. As a team we shift the paradigm.

Entering research to tap the zero-point energy is playing what I call the Prometheus game. In mythology, Prometheus gave fire to man, and for this the Gods punished him for eternity. An inventor who believes he alone is giving "free energy" to mankind is a pawn in this game. If his invention is successful, he will find himself under attack and ridicule. How can one safely play and win the game?

The answer lies within your higher self. Ask yourself the following question: If you were an angel who had the knowledge to seed the discovery of free energy on planet earth, would you love this planet and its beings enough to share your gift without any reward or recognition? If you can answer yes, then you are a master of the Prometheus game and you will find, as I have, that wonderful, synchronistic events and experiences accrue that yield inspiration and guidance. For in actuality, you *are* that angel. Your higher self has manifested a clear channel, a communicating vehicle known as your physical self, to accomplish its purpose. The purpose is to launch a repeating experiment around our globe.

The master player is totally free and fulfilled. No one can steal what is freely given. No one can suppress what repeats everywhere. As the paradigm shifts, the special interests that went into suppressing the discovery will then produce enormous investment capital to develop it (i.e., if you can't lick them, join them). At this point the free energy industry will grow with the same rapidity as the computer industry, offering numerous opportunities for creative research and development within the new science. By uplifting the world, we all win.

This book is dedicated to inventors for they are the ones who change the world. If each inventor shares a small experiment to be freely repeated by all others, the paradigm shifts. There is tremendous joy and fulfillment for those who freely share to uplift an entire planet. I invite you to join and experience the joyful transformation, for we are the creators who change the world.

ABOUT THE AUTHOR

Moray B. King, B.S. Electrical Engineering, M.S. Systems Engineering from the University of Pennsylvania, is currently a scientist at Eyring Research Inc., Provo, Utah. There he has co-developed an automated, broadband, antenna test system for field pattern testing of large HF antennas. As an advocation his main research interest involves supporting, with the standard physics literature, the speculation that a zero-point energy coherence can be induced by technological means. To encourage experimental research, he has given numerous presentations on this topic over the past 14 years.

THE A.T. FACTOR

A Scientists Encounter with UFOs: Piece For A Jigsaw Part 3

by Leonard Cramp

British aerospace engineer Cramp began much of the scientific anti-gravity and UFO propulsion analysis back in 1955 with his landmark book *Space, Gravity & the Flying Saucer* (out-of-print and rare). His next books (available from Adventures Unlimited) *UFOs & Anti-Gravity: Piece for a Jig-Saw* and *The Cosmic Matrix: Piece for a Jig-Saw Part 2* began Cramp's in depth look into gravity control, free-energy, and the interlocking web of energy that pervades the universe. In this final book, Cramp brings to a close his detailed and controversial study of UFOs and Anti-Gravity.

324 PAGES. 6X9 PAPERBACK. ILLUSTRATED. BIBLIOGRAPHY. INDEX. $16.95. CODE: ATF

COSMIC MATRIX

Piece for a Jig-Saw, Part Two

by Leonard G. Cramp

Leonard G. Cramp, a British aerospace engineer, wrote his first book *Space Gravity and the Flying Saucer* in 1954. Cosmic Matrix is the long-awaited sequel to his 1966 book *UFOs & Anti-Gravity: Piece for a Jig-Saw.* Cramp has had a long history of examining UFO phenomena and has concluded that UFOs use the highest possible aeronautic science to move in the way they do. Cramp examines anti-gravity effects and theorizes that this super-science used by the craft—described in detail in the book—can lift mankind into a new level of technology, transportation and understanding of the universe. The book takes a close look at gravity control, time travel, and the interlocking web of energy between all planets in our solar system with Leonard's unique technical diagrams. A fantastic voyage into the present and future!

364 PAGES. 6X9 PAPERBACK. ILLUSTRATED. BIBLIOGRAPHY. $16.00. CODE: CMX

UFOS AND ANTI-GRAVITY

Piece For A Jig-Saw

by Leonard G. Cramp

Leonard G. Cramp's 1966 classic book on flying saucer propulsion and suppressed technology is a highly technical look at the UFO phenomena by a trained scientist. Cramp first introduces the idea of 'anti-gravity' and introduces us to the various theories of gravitation. He then examines the technology necessary to build a flying saucer and examines in great detail the technical aspects of such a craft. Cramp's book is a wealth of material and diagrams on flying saucers, anti-gravity, suppressed technology, G-fields and UFOs. Chapters include Crossroads of Aerodymanics, Aerodynamic Saucers, Limitations of Rocketry, Gravitation and the Ether, Gravitational Spaceships, G-Field Lift Effects, The Bi-Field Theory, VTOL and Hovercraft, Analysis of UFO photos, more.

388 PAGES. 6X9 PAPERBACK. ILLUSTRATED. $16.95. CODE: UAG

THE TESLA PAPERS

Nikola Tesla on Free Energy & Wireless Transmission of Power

by Nikola Tesla, edited by David Hatcher Childress

David Hatcher Childress takes us into the incredible world of Nikola Tesla and his amazing inventions. Tesla's rare article "The Problem of Increasing Human Energy with Special Reference to the Harnessing of the Sun's Energy" is included. This lengthy article was originally published in the June 1900 issue of *The Century Illustrated Monthly Magazine* and it was the outline for Tesla's master blueprint for the world. Tesla's fantastic vision of the future, including wireless power, anti-gravity, free energy and highly advanced solar power. Also included are some of the papers, patents and material collected on Tesla at the Colorado Springs Tesla Symposiums, including papers on: •The Secret History of Wireless Transmission •Tesla and the Magnifying Transmitter •Design and Construction of a Half-Wave Tesla Coil •Electrostatics: A Key to Free Energy •Progress in Zero-Point Energy Research •Electromagnetic Energy from Antennas to Atoms •Tesla's Particle Beam Technology •Fundamental Excitatory Modes of the Earth-Ionosphere Cavity

325 PAGES. 8X10 PAPERBACK. ILLUSTRATED. $16.95. CODE: TTP

THE FANTASTIC INVENTIONS OF NIKOLA TESLA

by Nikola Tesla with additional material by David Hatcher Childress

This book is a readable compendium of patents, diagrams, photos and explanations of the many incredible inventions of the originator of the modern era of electrification. In Tesla's own words are such topics as wireless transmission of power, death rays, and radio-controlled airships. In addition, rare material on German bases in Antarctica and South America, and a secret city built at a remote jungle site in South America by one of Tesla's students, Guglielmo Marconi. Marconi's secret group claims to have built flying saucers in the 1940s and to have gone to Mars in the early 1950s! Incredible photos of these Tesla craft are included. The Ancient Atlantean system of broadcasting energy through a grid system of obelisks and pyramids is discussed, and a fascinating concept comes out of one chapter: that Egyptian engineers had to wear protective metal head-shields while in these power plants, hence the Egyptian Pharoah's head covering as well as the Face on Mars! •His plan to transmit free electricity into the atmosphere. •How electrical devices would work using only small antennas. •Why unlimited power could be utilized anywhere on earth. •How radio and radar technology can be used as death-ray weapons in Star Wars.

342 PAGES. 6X9 PAPERBACK. ILLUSTRATED. $16.95. CODE: FINT

QUEST FOR ZERO-POINT ENERGY

Engineering Principles for "Free Energy"

by Moray B. King

King expands, with diagrams, on how free energy and anti-gravity are possible. The theories of zero point energy maintain there are tremendous fluctuations of electrical field energy embedded within the fabric of space. King explains the following topics: Tapping the Zero-Point Energy as an Energy Source; Fundamentals of a Zero-Point Energy Technology; Vacuum Energy Vortices; The Super Tube; Charge Clusters: The Basis of Zero-Point Energy Inventions; Vortex Filaments, Torsion Fields and the Zero-Point Energy; Transforming the Planet with a Zero-Point Energy Experiment; Dual Vortex Forms: The Key to a Large Zero-Point Energy Coherence. Packed with diagrams, patents and photos. With power shortages now a daily reality in many parts of the world, this book offers a fresh approach very rarely mentioned in the mainstream media.

224 PAGES. 6X9 PAPERBACK. ILLUSTRATED. $14.95. CODE: QZPE

TAPPING THE ZERO POINT ENERGY

Free Energy & Anti-Gravity in Today's Physics

by Moray B. King

King explains how free energy and anti-gravity are possible. The theories of the zero point energy maintain there are tremendous fluctuations of electrical field energy imbedded within the fabric of space. This book tells how, in the 1930s, inventor T. Henry Moray could produce a fifty kilowatt "free energy" machine; how an electrified plasma vortex creates anti-gravity; how the Pons/Fleischmann "cold fusion" experiment could produce tremendous heat without fusion; and how certain experiments might produce a gravitational anomaly.

180 PAGES. 5X8 PAPERBACK. ILLUSTRATED. $12.95. CODE: TAP

THE FREE-ENERGY DEVICE HANDBOOK

A Compilation of Patents and Reports

by David Hatcher Childress

A large-format compilation of various patents, papers, descriptions and diagrams concerning free-energy devices and systems. *The Free-Energy Device Handbook* is a visual tool for experimenters and researchers into magnetic motors and other "over-unity" devices. With chapters on the Adams Motor, the Hans Coler Generator, cold fusion, superconductors, "N" machines, space-energy generators, Nikola Tesla, T. Townsend Brown, and the latest in free-energy devices. Packed with photos, technical diagrams, patents and fascinating information, this book belongs on every science shelf. With energy and profit being a major political reason for fighting various wars, free-energy devices, if ever allowed to be mass distributed to consumers, could change the world! Get your copy now before the Department of Energy bans this book!

292 PAGES. 8X10 PAPERBACK. ILLUSTRATED. BIBLIOGRAPHY. $16.95. CODE: FEH

ATLANTIS & THE POWER SYSTEM OF THE GODS

Mercury Vortex Generators & the Power System of Atlantis

by David Hatcher Childress and Bill Clendenon

Clendenon takes on an unusual voyage into the world of ancient flying vehicles, strange personal UFO sightings, a meeting with a "Man In Black" and then to a centuries-old library in India where he got his ideas for the diagrams of mercury vortex engines. The second part of the book is Childress' fascinating analysis of Nikola Tesla's broadcast system in light of Edgar Cayce's "Terrible Crystal" and the obelisks of ancient Egypt and Ethiopia. Includes: Atlantis and its crystal power towers that broadcast energy; how these incredible power stations may still exist today; inventor Nikola Tesla's nearly identical system of power transmission; Mercury Proton Gyros and mercury vortex propulsion; more. Richly illustrated, and packed with evidence that Atlantis not only existed—it had a world-wide energy system more sophisticated than ours today.

246 PAGES. 6X9 PAPERBACK. ILLUSTRATED. $15.95. CODE: APSG

THE TIME TRAVEL HANDBOOK

A Manual of Practical Teleportation & Time Travel

edited by David Hatcher Childress

In the tradition of *The Anti-Gravity Handbook* and *The Free-Energy Device Handbook*, science and UFO author David Hatcher Childress takes us into the weird world of time travel and teleportation. Not just a whacked-out look at science fiction, this book is an authoritative chronicling of real-life time travel experiments, teleportation devices and more. *The Time Travel Handbook* takes the reader beyond the government experiments and deep into the uncharted territory of early time travellers such as Nikola Tesla and Guglielmo Marconi and their alleged time travel experiments, as well as the Wilson Brothers of EMI and their connection to the Philadelphia Experiment—the U.S. Navy's forays into invisibility, time travel, and teleportation. Childress looks into the claims of time travelling individuals, and investigates the unusual claim that the pyramids on Mars were built in the future and sent back in time. A highly visual, large format book, with patents, photos and schematics. Be the first on your block to build your own time travel device!

316 PAGES. 7X10 PAPERBACK. ILLUSTRATED. $16.95. CODE: TTH

MAN-MADE UFOS 1944—1994

Fifty Years of Suppression

by Renato Vesco & David Hatcher Childress

A comprehensive look at the early "flying saucer" technology of Nazi Germany and the genesis of man-made UFOs. This book takes us from the work of captured German scientists to escaped battalions of Germans, secret communities in South America and Antarctica to todays state-of-the-art "Dreamland" flying machines. Heavily illustrated, this astonishing book blows the lid off the "government UFO conspiracy" and explains with technical diagrams the technology involved. Examined in detail are secret underground airfields and factories; German secret weapons; "suction" aircraft; the origin of NASA; gyroscopic stabilizers and engines; the secret Marconi aircraft factory in South America; and more.

318 PAGES. 6X9 PAPERBACK. ILLUSTRATED. FOOTNOTES. $18.95. CODE: MMU

THE ANTI-GRAVITY HANDBOOK

edited by David Hatcher Childress, with Nikola Tesla, T.B. Paulicki, Bruce Cathie, Albert Einstein and others

The new expanded compilation of material on Anti-Gravity, Free Energy, Flying Saucer Propulsion, UFOs, Suppressed Technology, NASA Cover-ups and more. Highly illustrated with patents, technical illustrations and photos. This revised and expanded edition has more material, including photos of Area 51, Nevada, the government's secret testing facility. This classic on weird science is back in a 90s format!

- **How to build a flying saucer.**
- **Arthur C. Clarke on Anti-Gravity.**
- **Crystals and their role in levitation.**
- **Secret government research and development.**
- **Bruce Cathie's Anti-Gravity Equation.**
- **NASA, the Moon and Anti-Gravity.**

230 PAGES. 7X10 PAPERBACK. ILLUSTRATED. $14.95. CODE: AGH

ANTI–GRAVITY & THE WORLD GRID

Is the earth surrounded by an intricate electromagnetic grid network offering free energy? This compilation of material on ley lines and world power points contains chapters on the geography, mathematics, and light harmonics of the earth grid. Learn the purpose of ley lines and ancient megalithic structures located on the grid. Discover how the grid made the Philadelphia Experiment possible. Explore the Coral Castle and many other mysteries, including acoustic levitation, Tesla Shields and scalar wave weaponry. Browse through the section on anti-gravity patents, and research resources.

274 PAGES. 7X10 PAPERBACK. ILLUSTRATED. $14.95. CODE: AGW

ANTI–GRAVITY & THE UNIFIED FIELD

edited by David Hatcher Childress

Is Einstein's Unified Field Theory the answer to all of our energy problems? Explored in this compilation of material is how gravity, electricity and magnetism manifest from a unified field around us. Why artificial gravity is possible; secrets of UFO propulsion; free energy; Nikola Tesla and anti-gravity airships of the 20s and 30s; flying saucers as superconducting whirls of plasma; anti-mass generators; vortex propulsion; suppressed technology; government cover-ups; gravitational pulse drive; spacecraft & more.

240 PAGES. 7X10 PAPERBACK. ILLUSTRATED. $14.95. CODE: AGU

THE GIZA DEATH STAR

The Paleophysics of the Great Pyramid & the Military Complex at Giza

by Joseph P. Farrell

Physicist Joseph Farrell's amazing book on the secrets of Great Pyramid of Giza. *The Giza Death Star* starts where British engineer Christopher Dunn leaves off in his 1998 book, *The Giza Power Plant*. Was the Giza complex part of a military installation over 10,000 years ago? Chapters include: An Archaeology of Mass Destruction, Thoth and Theories; The Machine Hypothesis; Pythagoras, Plato, Planck, and the Pyramid; The Weapon Hypothesis; Encoded Harmonics of the Planck Units in the Great Pyramid; High Fregquency Direct Current "Impulse" Technology; The Grand Gallery and its Crystals: Gravito-acoustic Resonators; The Other Two Large Pyramids; the "Causeways," and the "Temples"; A Phase Conjugate Howitzer; Evidence of the Use of Weapons of Mass Destruction in Ancient Times; more.

290 PAGES. 6X9 PAPERBACK. ILLUSTRATED. $16.95. CODE: GDS

A HITCHHIKER'S GUIDE TO ARMAGEDDON

by David Hatcher Childress

With wit and humor, popular Lost Cities author David Hatcher Childress takes us around the world and back in his trippy finalé to the Lost Cities series. He's off on an adventure in search of the apocalypse and end times. Childress hits the road from the fortress of Megiddo, the legendary citadel in northern Israel where Armageddon is prophesied to start. Childress muses on the rise and fall of civilizations, and the forces that have shaped mankind over the millennia, including wars, invasions and cataclysms. He discusses the ancient Armageddons of the past, and chronicles recent Middle East developments and their ominous undertones. In the meantime, he becomes a cargo cult god on a remote island off New Guinea, gets dragged into the Kennedy Assassination by one of the "conspirators," investigates a strange power operating out of the Altai Mountains of Mongolia, and discovers how the Knights Templar and their offshoots have driven the world toward an epic battle centered around Jerusalem and the Middle East.

320 PAGES. 6X9 PAPERBACK. ILLUSTRATED. BIB. & INDEX. $16.95. CODE: HGA

TECHNOLOGY OF THE GODS

The Incredible Sciences of the Ancients

by David Hatcher Childress

Popular *Lost Cities* author David Hatcher Childress takes us into the amazing world of ancient technology, from computers in antiquity to the "flying machines of the gods." Childress looks at the technology that was allegedly used in Atlantis and the theory that the Great Pyramid of Egypt was originally a gigantic power station. He examines tales of ancient flight and the technology that it involved; how the ancients used electricity; megalithic building techniques; the use of crystal lenses and the fire from the gods; evidence of various high tech weapons in the past, including atomic weapons; ancient metallurgy and heavy machinery; the role of modern inventors such as Nikola Tesla in bringing ancient technology back into modern use; impossible artifacts; and more.

356 PAGES. 6X9 PAPERBACK. ILLUSTRATED. BIBLIOGRAPHY. $16.95. CODE: TGOD

One Adventure Place
P.O. Box 74
Kempton, Illinois 60946
United States of America
Tel.: 815-253-6390 • Fax: 815-253-6300
Email: auphq@frontiernet.net
http://www.adventuresunlimitedpress.com
or www.adventuresunlimited.nl

ORDERING INSTRUCTIONS

✓ Remit by USD$ Check, Money Order or Credit Card
✓ Visa, Master Card, Discover & AmEx Accepted
✓ Prices May Change Without Notice
✓ 10% Discount for 3 or more Items

SHIPPING CHARGES

United States

✓ Postal Book Rate { $3.00 First Item, 50¢ Each Additional Item
✓ Priority Mail { $4.00 First Item, $2.00 Each Additional Item
✓ UPS { $5.00 First Item, $1.50 Each Additional Item

NOTE: UPS Delivery Available to Mainland USA Only

Canada

✓ Postal Book Rate { $6.00 First Item, $2.00 Each Additional Item
✓ Postal Air Mail { $8.00 First Item, $2.50 Each Additional Item
✓ Personal Checks or Bank Drafts MUST BE USD$ and Drawn on a US Bank
✓ Canadian Postal Money Orders OK
✓ Payment MUST BE USD$

All Other Countries

✓ Surface Delivery { $10.00 First Item, $4.00 Each Additional Item
✓ Postal Air Mail { $14.00 First Item, $5.00 Each Additional Item
✓ Payment MUST BE USD$
✓ Checks and Money Orders MUST BE USD$ and Drawn on a US Bank or branch.
✓ Add $5.00 for Air Mail Subscription to Future *Adventures Unlimited* Catalogs

SPECIAL NOTES

✓ RETAILERS: Standard Discounts Available
✓ BACKORDERS: We Backorder all Out-of-Stock Items Unless Otherwise Requested
✓ PRO FORMA INVOICES: Available on Request

✓ VIDEOS: NTSC Mode Only. Replacement only.
✓ For PAL mode videos contact our Europe office:
PO Box 48, 1600 AA Enkhuizen, The Netherlands

Please check: ☑

❐ This is my first order ❐ I have ordered before

Name
Address
City
State/Province Postal Code
Country
Phone day Evening
Fax

Item Code	Item Description	Qty	Total

Please check: ☑

Subtotal ➡
Less Discount-10% for 3 or more items ➡

❐ Postal-Surface — Balance ➡
❐ Postal-Air Mail (Priority in USA) — Illinois Residents 6.25% Sales Tax ➡
Previous Credit ➡
❐ UPS (Mainland USA only) — Shipping ➡
Total (check/MO in USD$ only) ➡

❐ Visa/MasterCard/Discover/Amex

Card Number

Expiration Date

10% Discount When You Order 3 or More Items!